Safdar Bashir

Compound Specific Stable Isotope Analysis

Safdar Bashir

Compound Specific Stable Isotope Analysis

Assessment of in situ transformation of Hexachlorocyclohexane using carbon stable isotope analysis

Südwestdeutscher Verlag für Hochschulschriften

Impressum / Imprint
Bibliografische Information der Deutschen Nationalbibliothek: Die Deutsche Nationalbibliothek verzeichnet diese Publikation in der Deutschen Nationalbibliografie; detaillierte bibliografische Daten sind im Internet über http://dnb.d-nb.de abrufbar.
Alle in diesem Buch genannten Marken und Produktnamen unterliegen warenzeichen-, marken- oder patentrechtlichem Schutz bzw. sind Warenzeichen oder eingetragene Warenzeichen der jeweiligen Inhaber. Die Wiedergabe von Marken, Produktnamen, Gebrauchsnamen, Handelsnamen, Warenbezeichnungen u.s.w. in diesem Werk berechtigt auch ohne besondere Kennzeichnung nicht zu der Annahme, dass solche Namen im Sinne der Warenzeichen- und Markenschutzgesetzgebung als frei zu betrachten wären und daher von jedermann benutzt werden dürften.

Bibliographic information published by the Deutsche Nationalbibliothek: The Deutsche Nationalbibliothek lists this publication in the Deutsche Nationalbibliografie; detailed bibliographic data are available in the Internet at http://dnb.d-nb.de.
Any brand names and product names mentioned in this book are subject to trademark, brand or patent protection and are trademarks or registered trademarks of their respective holders. The use of brand names, product names, common names, trade names, product descriptions etc. even without a particular marking in this works is in no way to be construed to mean that such names may be regarded as unrestricted in respect of trademark and brand protection legislation and could thus be used by anyone.

Coverbild / Cover image: www.ingimage.com

Verlag / Publisher:
Südwestdeutscher Verlag für Hochschulschriften
ist ein Imprint der / is a trademark of
OmniScriptum GmbH & Co. KG
Heinrich-Böcking-Str. 6-8, 66121 Saarbrücken, Deutschland / Germany
Email: info@svh-verlag.de

Herstellung: siehe letzte Seite /
Printed at: see last page
ISBN: 978-3-8381-3921-0

Zugl. / Approved by: Tuebingen, Eberhard Karls Univerity, Diss,. 2014

Copyright © 2014 OmniScriptum GmbH & Co. KG
Alle Rechte vorbehalten. / All rights reserved. Saarbrücken 2014

Assessment of *in situ* transformation of Hexachlorocyclohexane using carbon stable isotope analysis (CSIA)

Assessment of *in situ* transformation of Hexachlorocyclohexane using carbon stable isotope analysis (CSIA)

By
Dr. Safdar Bashir (PhD)

Preface

The presented book with the title 'Assessment of *in situ* transformation of Hexachlorocyclohexane using carbon stable isotope analysis (CSIA)' was prepared by Safdar Bashir in the time from January 2010 to December 2013 under supervision of Dr. Ivonne Nijenhuis and Dr. Hans H. Richnow. Data generation was conducted in the Department of Isotope Biogeochemistry at the Helmholtz-Centre for Environmental Research – UFZ in Leipzig.

Parts of this book were already published or are under review in peer-reviewed journals. Note that figures and text passages in the result chapters are taken from those publications without further indication.

Abstract

The European environment agency listed around 250,000 contaminated field sites which need to be cleaned due to the hazardous effects they pose on human and ecosystem health. This number is expected to increase over the next years. Similar is the case of hexachlorocyclohexane (HCH) contamination and it is estimated that four to six million tons of various HCH materials have been dumped worldwide, which need an urgent removal from the environment. HCH can undergo degradation by microorganisms indigenous to the soil or groundwater. Therefore natural attenuation (NA), relying on the *in situ* biodegradation of pollutants is considered as a cost effective remediation strategy. However, it requires accurate monitoring techniques. Carbon stable isotope analysis (CSIA) is a powerful technique to provide information on the extent of degradation. α-HCH as many other organic components appear as a racemic mixture of enantiomers in the environment and enantiomer fraction (EF) can provide information on biodegradation. The combination of enantiomeric fraction (EF), CSIA and the enantiomer selective stable isotope analysis (ESIA) has potential for distinguishing transformation processes of contaminants *in situ*.

To validate the applicability of CSIA for HCH, reaction-specific carbon isotope enrichment factors (ε_c) were determined in laboratory experiments for HCH isomers during aerobic and anaerobic degradation and compared with relevant abiotic reactions. Bulk enrichment factors determined for aerobic degradation of α- and γ-HCH by two *Sphingobium* spp. with similar reaction mechanism were similar (ε_c = −1.0 to −1.6 ‰ for α-HCH and ε_c = −1.5 to −1.7 ‰ for γ-HCH). Carbon isotope fractionation for aerobic degradation was smaller (ε_c = −1.0 to −1.6 ‰) as compared to anaerobic biodegradation experiments with *Dehalococcoides* sp. (ε_c = −5.5 ± 0.8 ‰) and mixed cultures (ε_c = −3.1 ± 0.4 ‰). For the first time anaerobic HCH transformation coupled with growth of *Dehalococcoides mccartyi* strain 195 was reported. Furthermore, isomer and enantiomer selective stable isotope fractionation of α-HCH was analyzed during biotic and abiotic reactions. Enantio-selective transformation and carbon isotope fractionation of α-HCH enantiomers was observed only in biotic reference studies. The extent of carbon isotope fractionation in biotic and abiotic transformation was compared to analyze the mechanism of bond cleavage. The enrichment factors of individual enantiomers $\varepsilon_{enantiomer}$ allowed calculating an average enrichment factor in all cases which was identical with bulk enrichment factors ε_{bulk} showing the validity of the analytical approach.

The extent and variability of carbon stable isotope fractionation in all laboratory investigations validate the applicability of CSIA as tool to characterize transformation of HCH in the environment. Furthermore, the ESIA method can help to distinguish biotic and abiotic reactions. The ESIA approach has probably potential for tracing the fate of other chiral contaminants in the environment.

The evaluation of CSIA at a contaminated field site demonstrated its potential for the identification of source zone of HCH contaminations and estimation of the extent of degradation down gradient the source zones.

In short, this study provides a concept for studying the transformation of HCH in the environment. The use of ESIA provides a comprehensive assessment of *in situ* degradation of α-HCH but also offers bases for tracing the fate of other contaminants containing chiral isomer with respect to i) providing evidence of degradation, ii) distinguishing pathways and iii) quantifying degradation at contaminated field sites.

Zusammenfassung

Laut Europäischer Umweltagentur existieren allein in Europa ca. 250.000 Altlastengebiete, die eine große Gefahr für Mensch und Umwelt darstellen. Es wird davon ausgegangen, dass die Anzahl dieser kontaminierten Standorte in den nächsten Jahren weiter zunimmt. Hexachlorcyclohexan (HCH) ist mit seinen Isomeren eine dieser problematischen Verbindungen, welche an zahlreichen kontaminierten Standorten zu finden ist. Schätzungsweise vier bis sechs Millionen Tonnen HCH Abfälle sind weltweit präsent, was den dringenden Bedarf einer gesicherten Entsorgung dieser Kontamination verdeutlicht.

HCHs werden im Boden und Grundwasser von verschiedenen Mikroorganismen abgebaut. Natürliche Rückhalte- und Abbauprozesse (*Natural Attenuation* - NA), im Speziellen der natürliche mikrobielle Schadstoffabbau ist demnach eine kostengünstige und schonende Sanierungsstrategie. Dieses Sanierungskonzept erfordert jedoch Überwachungsmethoden, um den Fortschritt des Abbaus und dessen Effizienz bewerten zu können. Die Analyse stabiler Isotope (carbon stable isotope analysis - CSIA) hat sich dabei als eine leistungsstarke Methode zur Bewertung der Abbaurate erwiesen. Zahlreiche Verbindungen, wie z.B. α-HCH, kommen als Racemat in der Umwelt vor. Untersuchungen zur Isotopenfraktionierung von Enantiomerengemischen (enantiomeric fraction - EF) können zusätzlich Informationen über den biologischen Abbau dieser Verbindungen liefern. Eine sogenannte Enantiomeren-spezifische Analyse stabiler Isotopen (*enantiomer selective stable isotope analysis* - ESIA) hat demzufolge das Potential zur Identifizierung, Bewertung und Unterscheidung von verschiedenen *in situ* Abbauprozessen.

Für die Validierung der Anwendbarkeit von CSIA im HCH-Abbau wurden reaktionsspezifische Isotopen-Anreicherungsfaktoren (ε_c) in biotischen aeroben und anaeroben als auch relevanten abiotischen Abbauexperimenten ermittelt und miteinander verglichen.

Im aeroben Abbau von α- und γ-HCH durch zwei verschiedene *Sphingobium* Stämme wurden übereinstimmende Anreicherungsfaktoren (ε_c = −1.0 bis −1.6 ‰ für α-HCH und ε_c = −1.5 bis −1.7 ‰ für γ-HCH) ermittelt, die vermutlich durch relativ ähnliche Reaktionsmechanismen hervorgerufen werden. Die Anreicherungsfaktoren des anaeroben γ-HCH-Abbaus durch zwei *Dehalococcoides* Stämme ($\varepsilon_c = -5.5 \pm 0.8$ ‰) und einer Mischkultur ($\varepsilon_c = -3.1 \pm 0.4$ ‰) waren dagegen signifikant höher im Vergleich zum aeroben Abbau. Im Rahmen dieser Arbeit konnte zudem erstmalig ein wachstumsabhängiger anaerober HCH-Abbau durch *Dehalococcoides mccartyi* Stamm 195 gezeigt werden.

Des Weiteren wurden die Methode zur ESIA entwickelt und am Beispiel des chiralen α-HCH Isomers in biotischen und abiotischen Abbaureaktionen untersucht. Dabei konnte eine enantiomerselektive Transformation sowie Isotopenfraktionierung ausschließlich in den biologischen Abbauexperimenten beobachtet werden. Basierend auf dem Vergleich der Isotopenfraktionierung des biotischen und abiotischen HCH-Abbaus konnten Aufschlüsse über den Mechanismus der Bindungsspaltung gegeben werden. Die Berechnung eines gemittelten Anreicherungsfaktors (ε_{bulk}) aus den Faktoren der einzelnen Enantiomere ($\varepsilon_{enantiomer}$) war identisch zu dem gemessenen Anreichungsfaktor (ε_{bulk}) ohne Enantiomerentrennung. Damit konnte die Analytik der ESIA auch erfolgreich validiert werden. Das Ausmaß und die Variabilität der Kohlenstofffraktionierung in den Laborstudien belegt die Anwendbarkeit der CSIA zur Untersuchung des HCH-Abbaus in der Umwelt. Zusätzlich kann mit Hilfe der ESIA Methode eine Differenzierung biologischer und chemischer Reaktionen ermittelt werden. Weiterhin ist die Verwendung der ESIA Methode zum Nachweis des biologischen Abbaus anderer organischer Schadstoffe denkbar. Die Evaluierung der CSIA am kontaminierten Feldstandort verdeutlicht das Potential zur Identifizierung von HCH-Kontaminationsquellen und erlaubt die Abschätzung der Abbaurate entlang der Grundwasserströmung.

Zusammenfassend bietet diese Arbeit ein neues analytisches Konzept zur Abschätzung des *in situ* Abbaus von HCH an. Durch die Verknüpfung von EF und CSIA zur Entwicklung der ESIA Methode wurde ein umfassender Ansatz zum *in situ* Abbau von α-HCH vorgestellt. Dieser kann als Model für weitere Schadstoffe mit chiralen Isomeren bezüglich (i) eines Nachweis des Abbaus, (ii) der Unterscheidung verschiedener Abbauwege und (iii) einer Quantifizierung des *in situ* Abbaus dienen.

Acknowledgement

The data presented in this book has been generated at the Helmholtz Centre for Environmental Research - UFZ in the department of Isotope biogeochemistry and would not have been possible without the support, guidance and help of many people who in one way or another contributed to the preparation and completion of this doctoral book.

First and foremost I would like to thank my supervisor Dr. Hans H. Richnow for giving me the opportunity to do my PhD in the departement of Isotope Biogeochemistry as well as for providing valuable advice throughout the four years .

The second person on the list is of course Ivonne Nijenhuis, my supervisor, without whom all this would simply not have been possible. Despite the misunderstandings, your adaptation to my somewhat unconventional schedules and your ease to prioritize things, almost at the opposite of my skills, were truly admired. Your straight-to-the-point comments, quick nights and week-ends feedbacks were greatly appreciated. Your strong support and encouragements really lifted me up. In the worse moments you said the words I needed to hear, and I could never thank you enough for that. I learned...a lot... and if you did not benefit from it, I can assure you that my career will!

Anko Fischer the true isotope master! Your calmness and light control have been stress depleting factors. You are an inexhaustible source of wisdom which spreads motivation and knowledge enrichment. You were the stable soul beside the radioactive character.

I am also very grateful to Prof. Dr. Peter Grathwohl who supervised my work on behalf of the faculty of Applied Geosciences of the University of Tübingen.

On the road, one meets people that are true gems. Kevin, Sara, Jan, Conrad, Alex, Kristina, Ning, Tania, Rizwan, Diana and Marie are few of them. You were (and are!) a constant source of inspiration and admiration. You were the ultimate mentor! Your patience and expertise have been pushing me forward. Thank you for all the productive discussions, your thorough feedbacks, and all the good vibes you sent.

Many thanks go to my UFZ colleagues, especially the ones from the Isotope biogeochemistry Department. There was always a friendly and good atmosphere, e.g. during the numerous "coffe&cake sessions". I wish to acknowledge all the members of the ISOBIO team for their contribution to a most pleasant working environment, the nice and/or crazy moments we have had together and their support. I particularly would like to express my deep gratitude to Matthias Gehre, Ursula Günther, Ines Mäusezahl, Kerstin Ethner, Stefanie Hinke, and Falk for their organizational support in the labs and to Ms. Liane Paul for her help in administrative stuff.

Finally, I wish to express my love and gratitude to my family. You always supported, encouraged and believed in me during my studies and PhD time. Without you, I would not have been able to keep on doing what I did. Thank you. Also, I would like to thank my grandpa who always wanted to see me handing in my book – I hope grandpa you see me writing these last lines of my book and I feel sad to think of not being around me to tell you that your prays worked.

In the end, The University of Agriculture Faisalabad, Pakistan and HIGRADE are acknowledged for the financial support.

List of abbreviations

CB	Chlorobenzene
CSIA	Carbon stable isotope analysis
DCB	Dichlorobenzene
DNA	Deoxyribonucleic acid
Ee	Activation energy
EF	Enantiomer fractionation
ESIA	Enantiomer-selective stable isotope analysis
Eq.	Equation
GC-C-IRMS	Gas chromatography-combustion-isotope ratio mass spectrometry
GC-FID	Gas chromatography-Flame Ionisation Detector
GC-IRMS	Gas chromatography-Isotope Ratio Mass Spectrometry
GC-MS	Gas chromatography-Mass Spectroscopy
HCH	Hexachlorocyclohexane
IAEA	International Atomic Energy Agency
MCB	Monochlorobenzene
MNA	Monitored Natural Attenuation
NA	Natural Attenuation
n.a.	Not assessed
n.d.	Not determined
PCE	Tetrachloroethene / Tetrachloroethylene
PCR	Polymerase chain reaction
POP	Persistent organic pollutants
RDH	Reductive dehalogenation
RNA	Ribonucleic acid
rRNA	Ribosomal ribonucleic acid
SI	Supporting Information
SIFA	Stable isotope fractionation analysis
SIP	Stable isotope probing
TCB	Trichlorobenzene
TCE	Trichloroethene
TEA	Terminal electron acceptor
TiO_2	Titanium dioxide
UFZ	Helmholtz-Zentrum für Umweltforschung (Centre for Environmental Research)
US-EPA	United States Environmental Protection Agency
UV	Ultra violet
VC	Vinyl Chloride
V-PDB	Vienna PeeDee Belemnite
V-SMOW	Vienna Standard Mean Ocean Water
α	Isotope Fractionation Factor
$\delta^{13}C_{(x)}$	Carbon isotopic composition of a compound (x)
γ-PCCH	γ-pentachlorocyclohexene
ε	Enrichment factor

Content

1 Introduction 1

1.1 Hexachlorocyclohexane isomers 1

1.2 Properties and persistence 3

1.3 Contamination and risks for the environment 5

1.4 Transformation of HCH 7

    1.4.1 Abiotic processes 7

    1.4.2 Biotic processes 8

1.5 Contaminated site remediation strategies 11

1.6 Approaches for assessment of *in situ* biotransformation of HCH 12

1.7 CSIA application to monitor *in situ* biodegradation 13

    1.7.1 Fractionation of stable isotopes 13

    1.7.2 Stable isotope analysis of carbon and natural attenuation 14

1.8 Enantiomeric fraction 15

    1.8.1 Quantification of enantiomeric fractionation 16

1.9 Application of CSIA for HCH transformations 16

2 Objectives and strucrute of the book 27

3 Results 30

3.1 Enantioselective carbon stable isotope fractionation of hexachlorocyclohexane during aerobic biodegradation by *Sphingobium* spp. 31

    3.1.1 Introduction 32

    3.1.2 Material and methods 34

    3.1.3 Results and discussion 38

3.2 Anaerobic biotransformation of hexachlorocyclohexane isomers by *Dehalococcoides* spp. 55

    3.2.1 Introduction 57

    3.2.2 Material and methods 59

    3.2.3 Results and discussion 62

3.3 The application of compound specific isotope analysis to characterize abiotic reaction mechanisms of alpha-hexachlorocyclohexane. 76

    3.3.1 Introduction 77

    3.3.2 Material and methods 80

    3.3.3 Results and discussion 83

3.4 Evaluating degradation of hexachlorcyclohexane (HCH) isomers within a contaminated aquifer using compound specific stable carbon isotope analysis. 100

    3.4.1 Introduction 101

    3.4.2 Material and methods 104

    3.4.3 Results and discussion 108

4 Discussion 123

5 Summary & outlook 139

A Appendix 143

    A1 Supporting information chapter 3.1 144

    A2 Supporting information chapter 3.2 148

    A3 Supporting information chapter 3.3 150

    A4 Supporting information chapter 3.4 167

Chapter : 1 Introduction

Introduction

Introduction

1.1 Hexachlorocyclohexane isomers

Hexachlorocyclohexane (HCH) is one of the most widely produced and applied pesticides. It was originally synbooked in 1825 by Michael Faraday, however, its pesticidal property was discovered in 1942. Then commercial production of HCH by photochemical chlorination of benzene in the incidence of UV-light started at Imperial Chemical Industries Ltd. (Österreicher-Cunha et al., 2003). Technical-grade HCH comprises mainly five isomers bearing different stabilities: α- (60 to 70 %), β- (5 to 12 %), γ- (10 to 12 %), δ- (6 to 10 %) and ε- (3 to 4 %) along with traces of η- and θ-isomers (Vijgen, 2006). α-HCH is the sole chiral isomer with two enantiomers (+) α-HCH and (-) α-HCH (Li, 1999). The isomers of HCH vary in their spatial positions of the chlorine atoms around the cyclohexane ring which can be axial or equatorial (Figure 1). At first, all the isomers in the technical-grade HCH have been used intensively as pesticides; however, it was discovered afterwards that only the γ-isomer (Lindane) has insecticidal properties (Slade, 1945). Then developed countries started purification of the γ-isomer by treating the isomeric mixture of HCH with methanol or acetic acid and recrystallizing it to produce 99% pure γ-HCH (Lindane) (Willett et al., 1998).

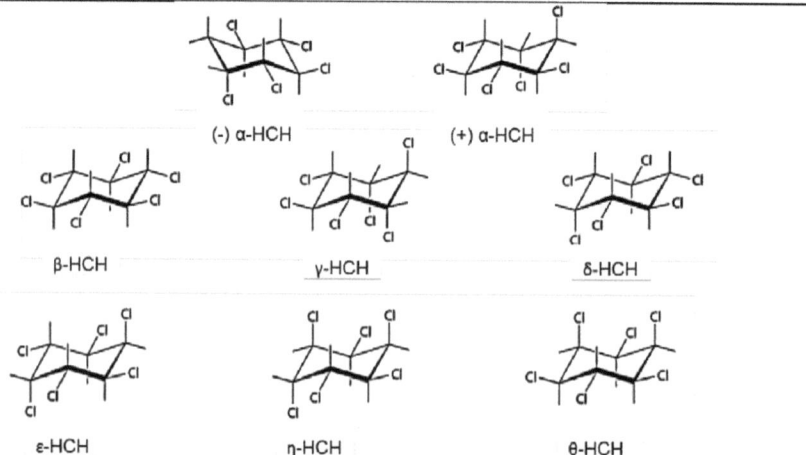

Figure 1: Axial and equatorial arrangements of chlorine atoms in the main HCH isomers (modified from Lal et al., 2010).

In late 1970s, the ill-famed pesticide dichlorodiphenyltrichloroethane (DDT) was banned and replaced with HCH which was conceived as an efficient alternative at that time because of its low cost and high effectiveness (De Paolis et al., 2013). HCH was used as a broad-spectrum insecticide for a variety of applications in agriculture, forestry and public health (Haugen et al., 1998). Typical utilizations were to protect vegetables, fruits, rice, in forestry production (e.g. Christmas trees), for seed treatment as well as to treat livestock and pets. Additionally, it was used as an insecticide in pharmaceuticals, lotions and shampoos. About 10 million tons of technical HCH have been used from 1940 to 1997 (Li, 1999) and around 382 thousand tons of technical HCH and 81 thousand tons of Lindane were used in Europe from 1970 to 1996 (Breivik et al., 1999). The persistency, toxicity and potential carcinogenicity of the other isomers have resulted in the ban of HCH usage in most of the countries (Deo et al., 1994; Vijgen et al., 2011).

1.2 Properties and persistence

The properties such as solubility, volatility, sorption and polarity of pesticides are mainly affected by the chemical structure which may contribute to their transport, persistence and biodegradability. A short summary of HCH properties is provided in Table 1. The persistence of each isomer is largely dependent on the alignment of the chlorine atoms on the cyclohexane ring. As a general rule, less chlorine atoms at equatorial position or more at axial position renders persistence to the molecule. For instance, β-HCH is the most persistent isomer (Deo et al., 1994; Johri et al., 1996; Phillips et al., 2005; Sahu et al., 1990) having all chlorine atoms at equatorial positions (eeeeee), which results in more stable structure. Less persistent α-, δ- and γ-isomers have chlorine atoms oriented in equatorial and axial positions as 'aaaaee', 'aeeeee' and 'aaaaee', in that order (Figure 1) (Phillips et al., 2005).

The presence of axial chlorine atoms is thought to provide available sites for enzymatic degradation. It is generally observed that γ- and α-HCH are more easily biodegraded than the δ-isomer, which has more equatorial chlorine atoms (Deo et al., 1994). Persistence of HCHs in soil, water and air is puzzling because of variations in the available data due to complex interactions in the environment, which can affect both abiotic and biotic degradation (Phillips et al., 2005).

Table 1: Properties of most abundant HCH isomers (Phillips et al., 2005).

	α-HCH	β-HCH	δ-HCH	γ-HCH
Boiling point (°C)	288 at 760 mm Hg [a]	60 at 0.5 mm Hg [a]	60 at 0.36 mm Hg [a]	323.4 at 760 mm Hg [a]
Melting point (°C)	159-160 [a,b] 158 [c]	314-315 [a] 309-310 [b]	141-142 [a] 138-139 [b]	112.5 [a,c,d] 112-113 [b]
Density (g cm^{-3})	1.87 at 20 °C [a]	1.89 at 20 °C [a,c]	NA [h]	1.89 at 19 °C [a]
Solubility in water (mg L^{-1})	10 [a]	5 [a]	10 [a]	0-17 [f] 7 [g]
Solubility in 100 g ethanol (mg)	1.8 [a]	1.1 [a]	24.4 [a]	6.4 [a,c]
Log K_{ow}	3.8 [a] 3.9±0.2 [b]	3.78 [a] 3.9±0.1 [b]	4.14 [a] 4.1±0.02 [b]	3.72 [a] 3.7±0.5 [b] 3.85 [d]
Log K_{oc}	3.57 [a]	3.57 [a]	3.8 [a]	3.57 [a] 3.04 [g]
Vapour pressure (mm Hg)	4.5 x 10^{-5} at 25 °C [a]	3.6 x 10^{-7} at 20 °C [a]	3.5 x 10^{-5} at 25 °C [a]	4.2 x 10^{-5} at 20 °C [a] 9.4 x 10^{-6} at 20 °C [c] 3.1 x 10^{-5} at 25 °C [d] 3.3 x 10^{-5} at 20-25 °C [g]

[a] (Dorsey, 2005)
[b] (Willett et al., 1998)
[c] CRC Handbook of Chemistry and Physics (Lide, 2004)
[d] (Bintein and Devillers, 1996).
[e] Merck Index.
[f] Although this range of solubility has been reported in the literature, the consent value appears to be closer to 7–10 mg L^{-1} (Phillips et al., 2005).
[g] (Paraiba and Spadotto, 2002)

HCH-isomers are persistent and can stay in soil for more than 11 years (Lichtenstein and Polivka, 1959). Reported half-lives for Lindane in soil range from 6 weeks to 260 days (Jury et al., 1983) to 2 years (Bintein and Devillers, 1996; Johri et al., 1996). Application of HCH isomers to a sandy loam soil field showed their persistence over a time period of 15 years (Stewart and Chisholm, 1971). Generally, half-life depends on the method of application, initial concentrations, soil properties and environmental conditions (Phillips et al., 2005). The α-isomer is the only chiral isomer of HCH. It can exist in two enantiomer forms (+) α-HCH and (-) α-HCH as shown in Figure 1.

1.3 Contamination and risks for the environment

Anthropogenic activities have introduced toxic chemicals into the soil and groundwater throughout the world. In 2006, the European Environment Agency stated around 250,000 contaminated field sites which urgently need to be cleaned to avoid hazards they pose on human and ecosystem health. This number is expected to increase over the next years. A similar situation exists in the United States where about 294,000 contaminated field sites are known (Bombach et al., 2010). These contaminated field sites may contain a variety of contaminants, which may range from simple organic contaminants to persistent organic pollutants (POPs) including HCH contamination. It is estimated that four to six million tons of various HCH materials have been dumped worldwide. The order of HCH containing waste is similar in range to the sum of other dumped persistent organic pollutants (POPs) defined by the Stockholm Convention (Weber et al., 2008). In 2009, the UNEP Stockholm Convention has listed the α, β and γ isomers of HCH as POPs (Vijgen et al., 2011).

HCH contamination in the environment is of two major types. One involves the point source contamination of very high concentrations because of improper waste disposal in extreme cases and open air stock piling of waste isomers from Lindane production, because the production of one ton of Lindane creates 8 to 10 tons of HCH waste, which was dumped into the environment in an inappropriate manner (Vijgen, 2006; Weber et al., 2008). Heavily contaminated point sources are shown in Figure 2. A second type of contamination is related to diffuse contamination of the environment with lower HCH concentrations due to either dispersion from the stockpiles or the application of the insecticide (Lal et al., 2010).

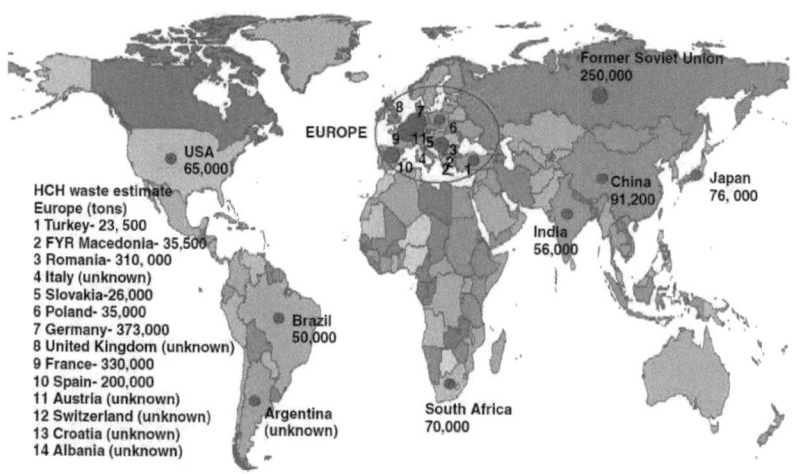

Figure 2: Preliminary estimates on quantities of stored/deposited HCH waste isomers present in different countries (taken from Vijgen et al., 2011).

HCH residues at many sites have been reported to percolate into the soil and groundwater (Ricking and Schwarzbauer, 2008; Wycisk et al., 2013). Due to worldwide spread of HCH by various transport mechanisms and diffuse contamination, HCH residues have been found in glaciers as well (Ørbæk et al., 2004). The most abundant organochlorine in arctic air, freshwaters and the Arctic Ocean is HCH, which symbolizes the extreme dispersion of HCH in the environment (Hinckley et al., 1991).

Humans can be exposed to HCH isomers through various means, which include such as ingestion of contaminated water or food, absorption through skin or by inhalation. The α-, β- and γ-isomers of HCH mostly act as sedative of the nervous system (Nagata et al., 1999). Carcinogenic effects of HCH isomers in mice were also reported (Sagelsdorff et al., 1983). HCHs have been additionally reported as endocrine disrupters in which γ- and β-HCH have been shown to have weak estrogenic activity, and together with the α- and the δ-isomer can interfere with steroid synbook (Olivero-Verbel et al., 2011). The β-HCH is toxicologically most significant due to its high persistence in mammalian tissues and its estrogenic effects in mammalian cells and fish (Willett et al., 1998). β-HCH is classified as a xenobiotic that creates estrogen-like effects through non-classical mechanisms and may be of importance with regard to breast and uterine cancer risks (Steinmetz et al., 1996). The bio-concentration factor of β-HCH in human fat is nearly 30 times greater than that of γ-HCH (Geyer et al., 1987). The average half-life of β-HCH in the blood is 72 years (Jung et al., 1997) as compared to only 1 day for γ-HCH (Feldmann and Maibach, 1974).

1.4 Transformation of HCH

For the successful application of remediation technologies and monitoring natural attenuation, it is important to understand the processes governing transformation of HCH. The details of major HCH transformations i.e. abiotic and biotic (aerobic and anaerobic) are given below.

1.4.1 Abiotic processes

A number of studies have already well documented HCHs treatment by various chemical technologies. Direct photolysis by short-wavelength UV irradiation (100-280 nm) may directly transform α-, β- and δ-isomers of HCH (Hamada et al., 1982). However it is unlikely that short-wavelength UV light act on HCHs at the Earth's surface level because this part of the light is completely filtered by the atmosphere. Indirect photolysis such as H_2O_2-assisted UV irradiation (photo-Fenton process) can transform γ-HCH via powerful OH radical or via TiO_2-enhanced photocatalysis (EPA, 1999; Hatzinger et al., 2013; Wiegert, 2013). The hydrolysis mechanisms and pH dependence of reaction rate constants for γ-HCH were discussed and a dehydrochlorination pathway is proposed (Ngabe et al., 1993). In addition, distinctly different rates and pathways were observed in the presence of solid iron sulphide (Bockelmann et al., 2003). Direct reduction of γ-HCH at carbon cathodes has been explored by means of cyclic voltammetry and controlled-potential electrolysis (EPA, 1999). Nanoscale zero-valent iron for treatment of contaminated ground water or soil by HCHs represent a new generation of environmental remediation technologies that could provide cost-effective solutions to some of the most challenging environmental clean-up problems (Elliott et al., 2009; Singh et al., 2011; Wiedemeier, 1999; Zhang, 2003). But chemical transformation processes are not always cost effective and often toxic to *in situ* microbes and could result in partial transformation of contaminants

(Norris et al., 1995; Pardieck et al., 1992). Thus chemical technologies might be not sufficient for a considerable removal of HCH from the environment. In this case an important mechanism which substantially removes HCH from the environment is biodegradation.

1.4.2 Biotic processes

Biotic processes are responsible for the significant removal of contaminants and are a cost effective method to substantially remove contaminants from the contaminated site. Bacteria are usually vital agents in most biotic transformation processes. Most of synthetic organic chemicals in soils and water are transformed to inorganic products by bacterial metabolism and co-metabolism. These microbial processes may lead to removal of contaminants from the environment. The mode of degradation of various halogenated contaminants with isolated pure cultures and their genetic characterization has been investigated intensively under oxic and anoxic conditions (Anderson et al., 1993; Baker and Herson, 1994; Lal et al., 2010; Phillips et al., 2005).

1.4.2.1 Oxic conditions

In late 1980s, the aerobic degradation of HCHs was reported for the first time (Bachmann et al., 1988a; Sahu et al., 1993; Sahu et al., 1995). Most HCH degrading aerobic bacteria belong to the family *Sphingomonadaceae* (Lal et al., 2010) however, other bacterial strains were reported to degrade different HCH-isomers (Bachmann et al., 1988b; Manickam et al., 2006). The first HCH degrading isolate was classified as *Pseudomonas paucimobilis* strain UT26 and derived from an upland experimental field in Japan where γ-HCH had been applied once a year over 12 years (Senoo and Wada, 1989). Strain UT26 uses γ-HCH as a sole carbon and energy source. Further phylogenetic related strains were isolated from a rice field from India (Sahu et al., 1990) and from river Rhine in France (Thomas et al., 1996), respectively. Later 16S rRNA gene sequence analysis revealed that all three strains are distinct species of the genus *Sphingobium* and thus named as *Sphingobium japonicum* UT26, *S. indicum* strain B90A and *S. francense* Sp+, respectively (Pal et al., 2005). During the last decade numerous strains of this genus were isolated from HCH-contaminated sites in China (Ma et al., 2005), Japan (Ito et al., 2007), India (Dadhwal et al., 2009), Germany (Böltner et al., 2005) and Spain (Mohn et al., 2006). *S. japonicum* strain

UT26 and *S. indicum* strain B90A were investigated intensively for HCH degradation, metabolites formation, expression of responsible genes and enzymes during transformation (Lal et al., 2010).

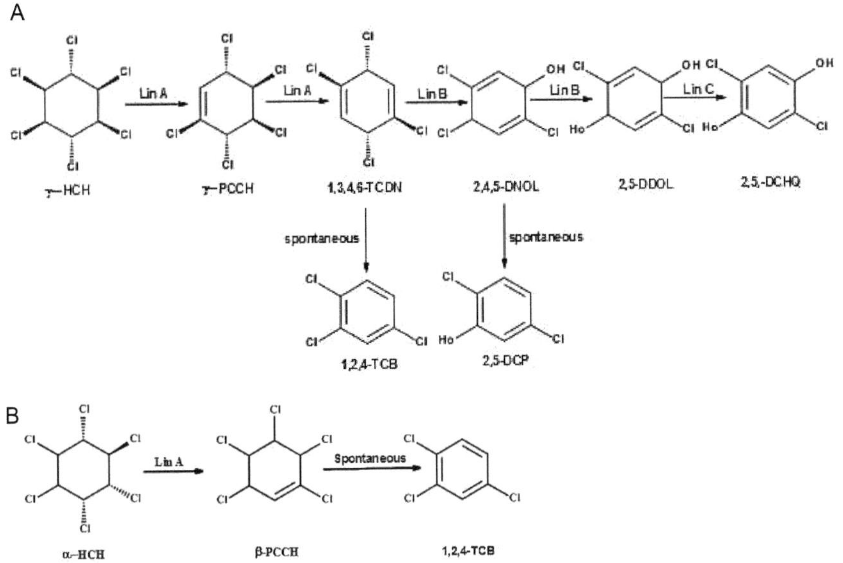

Figure 3: Aerobic biodegradation pathways for γ-HCH (A) and α-HCH (B) by bacterial strains *S. indicum* strain B90A and *S. japonicum* strain UT26, respectively. (modified from Lal et al., 2010)

The aerobic transformation of γ-HCH is characterized by two initial dehydrochlorination reactions producing the assumed product 1,3,4,6-tetrachloro-1,4-cyclohexadiene (1,3,4,6-TCDN) through the intermediate γ-pentachlorocyclohexene (γ-PCCH) (Imai et al., 1991; Nagasawa et al., 1993), whereby the other end products 1,2,4-trichlorobenzene (1,2,4-TCB) and 2,5-dichlopenol (2,5-DCP) are formed depending on the different reaction routes (Geueke et al., 2013) (Figure 3 A). The α-HCH biotransformation proceeds similarly via dehydrochlorination, β-PCCH is produced as the first intermediate which is then converted into 1,2,4-TCB (Suar et al., 2005) (Figure 3 B). In both reaction pathways the first reaction steps are catalysed by LinA (enzyme necessary for aerobic transformation of α and γ-HCH). For δ-HCH and β-HCH a similar transformation mechanism (hydrolytic dechlorination) is reported. Here, the first transformation steps are catalysed by LinB resulting in 2,3,4,5,6-pentachlorocyclohexanol (PCHL) as intermediate. Recently, two *Arthrobacter* strains, which were not directly isolated from a HCH-polluted site, showed the potential to grow in a mineral salt medium containing α-, β-, or γ-HCH as sole source of carbon (De Paolis et al., 2013).

1.4.2.2 Anoxic conditions

Initially it was believed that biodegradation of HCH only occurs under anaerobic conditions. Due to the absence of a bacterial model for HCH degradation, rat liver microcosms with enriched cytochrome P-450 were initially used to study the degradation and relative rates of dechlorination of α-, β- and γ-HCH under anoxic conditions (Beurskens et al., 1991). Metabolites detected during anaerobic degradation of α-, β- and γ-HCH were δ-3,4,5,6-tetrachlorocyclohexene (TCCH) and

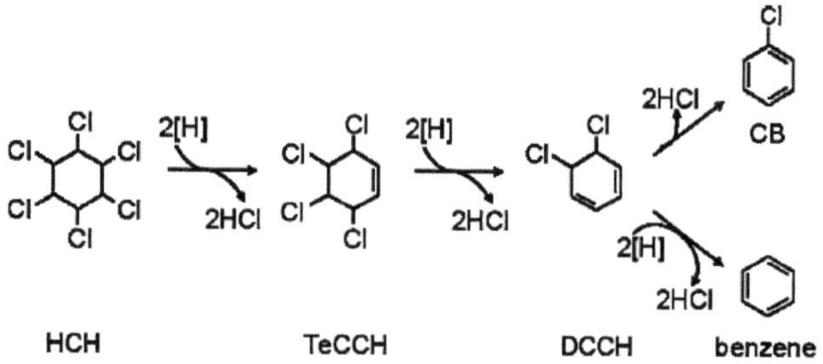

Figure 4: Proposed anaerobic biodegradation pathway of HCH (taken from Mehboob et al., 2013)

monochlorobenzene (MCB) (Figure 4). The degradation rates (γ-HCH>α-HCH>β-HCH) in this study were the same as found under anoxic environmental conditions (Heritage and MacRae, 1977; Macrae et al., 1969). All four main isomers of HCH were degraded faster in flooded soil than in sterilized soil and ^{14}C labelled CO_2 was produced from non-sterile flooded soils treated with ^{14}C labelled γ-HCH indicating mineralisation (MacRae et al., 1967). Anoxic studies clearly suggest that the anaerobic degradation of HCH isomers taking place through successive dechlorination and/or dehydrochlorination to produce benzene and/or chlorobenzene (Buser and Mueller, 1995; Doesburg et al., 2005; Van Liere et al., 2003) (Figure 4).

1.5 Contaminated site remediation strategies

Contaminated site remediation strategies vary depending on the site, extent of contamination, historical background of contamination, geographical characteristics and geochemical situations. Thus, reliable monitoring and precise risk assessment is needed to select the most suitable remediation technique. Some of the popular remediation technologies include bioremediation, mechanical soil aeration, chemical treatment, incineration, neutralization, phytoremediation, soil vapor extraction, soil washing, solidification/stabilization, solvent extraction or pump and treat (Sharma and Reddy, 2004; Wiegert, 2013).

However, most of the conventional remediation approaches are time consuming and expensive and lead often only to partial remediation. Since 1990s, natural attenuation (NA) has gained attention, because it relies on *in situ* biodegradation of the contaminants, without human involvement (Bombach et al., 2010; EPA, 1999). NA refers to the decrease of mass, volume, toxicity, mobility, and/or concentration of contaminants in soil or groundwater by naturally occurring processes. These processes can be physical, chemical or biological. Consequently, they include biodegradation, dispersion, dilution, sorption, volatilization, radioactive decay, chemical or biological stabilization, transformation, and destruction of contaminants (EPA, 1999; Wiedemeier, 1999). Biodegradation is one of the main processes, therefore, *in situ* biodegradation has been the topic of many investigations over the past decades (Schirmer et al., 2006; Voldner and Li, 1995; Wiedemeier, 1999).

In order to use NA as an effective remediation approach, some concerns must be taken into consideration including occurrence, efficiency and duration of biodegradation, so that the contaminant eradication takes place in a sensible time period (Bombach et al., 2010; Martin et al., 2003). Thus, robust and cost efficient monitoring tools are essential to validate the sustainability of NA (Bockelmann et al., 2003). Thus monitoring is essential to validate that NA works in a sustainable way and the term Monitored Natural Attenuation (MNA) is commonly used. Several methods have been established to evaluate NA at contaminated field sites, and more accurately to perfectly qualify and quantify the degradation processes. These methods include e.g. geochemical approaches, tracer tests, metabolite analysis, and microbial or molecular methods (Bombach et al., 2010; Schirmer et al., 2006). Concentration based estimations for biodegradation are frequently hindered by other physical processes, such as dilution and dispersion, which prevent creating a reliable mass balances. Therefore several methods are often combined to answer the questions and provide reliable clues for NA. However direct quantification of the amount of degradation and identification of the basic pathways is often not possible only by these methods (Bombach et al., 2010).

1.6 Approaches for assessment of *in situ* biotransformation of HCH

Considering the costs and technical difficulties linked to other conventional remediation techniques, currently, *in situ* biodegradation is considered as an effective approach for contaminants removal from the soil and water bodies. *In situ* degradation is the key process for NA strategies for managing contaminated sites (Schirmer et al., 2006). Only few studies demonstrated the stimulated transformation of HCH to intermediates in a full scale anaerobic *in situ* bioscreen, combined with an aerobic on site treatment to harmless end products (Langenhoff et al., 2013). However, complete characterization of site-specific biodegradation processes is essential to validate the effectiveness of *in situ* biodegradation of organic contaminants at a contaminated site and to investigate the efficiency of *in situ* remediation processes to replace conventional clean-up technologies (Pope et al., 2004). Recently, several methods or combination of various techniques have received great attention to verify the *in situ* biodegradation processes. In case of HCH there are several limitations e.g. the concentrations of HCH on most of the dump sites are too high to monitor concentration dependent biodegradation quantification. In case of application of molecular methods for assessing *in situ* transformation of HCH, several studies demonstrated successful application in oxic conditions because of the availability of well characterized bacterial strains (Lal et al., 2010) but the application of molecular approaches under anoxic conditions is limited because of lack of known and well characterized microorganisms. The lack of well characterized bacterial cultures is a bottleneck in studying the genetics and biochemistry of the anaerobic microbial HCH dechlorination process and monitoring natural attenuation with molecular methods (Mehboob et al., 2013). Enantiomer fractionation provides another indicator for *in situ* biodegradation of α-HCH and various studies showed the site specific enantioselective transformation of α-HCH enantiomers (Harner et al., 2000; Helm et al., 2000; Law et al., 2004). One of the most promising monitoring tools to characterize and assess contaminant sources and *in situ* degradation of organic contaminants is carbon stable isotope analysis (CSIA) (Aelion et al., 2010; Elsner, 2010; Hofstetter and Berg, 2011; Schirmer et al., 2006; Wijker et al., 2013).

1.7 CSIA application to monitor *in situ* biodegradation

Carbon stable isotope analysis (CSIA) has been shown to be an effective tool to evaluate *in situ* biodegradation processes (Swartjes, 2011). Stable isotope ratio measurements of specific compounds can be used to distinguish biodegradation from non-destructive processes affecting concentration only and has been reviewed extensively (Thullner et al., 2012). Isotope effects can also be used to explain reaction mechanisms and in combination with spatial/temporal data, can be applied *in situ* to estimate reaction rates. Moreover, in recent years, attempts have been made to use stable isotope fractionation data to discriminate different reaction pathways of contaminant degradation in groundwater (Feisthauer et al., 2012).

1.7.1 Fractionation of stable isotopes

Isotopes (Greek isos = "equal", topos = "site, place"; "at the same place" within the periodic table of elements) are atoms of one element with varying mass numbers due to different numbers of neutrons. Neutrons are neutral nuclear particles which decrease repulsive forces between protons. They are responsible for the variety of stable isotopes within nature. The effectiveness of stable isotopes for evaluating biodegradation is based on the fact that chemical bonds formed by a heavy isotope of an element typically are stronger than those formed by a light isotope of the same element in the same compound. Thus, when a specific bond is broken, molecules containing heavy isotopes of elements involved in the bond (or neighbouring to a lesser degree) generally react more slowly than molecules containing light isotopes of those elements (Figure 5) (Bigeleisen, 2004). This process is called as isotope fractionation and, when it is related to a unidirectional reaction (rather than a reversible phase change or other equilibrium process), as a kinetic isotope effect is

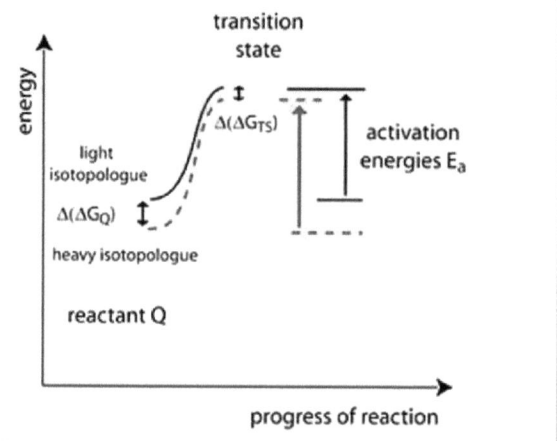

Figure 5: Principle of isotopic fractionation in microbially catalysed reactions. Lighter isotopologues require less activation energy (E_a) to achieve the transition state and will therefore react faster than heavy isotopologues. $\Delta(\Delta G_Q)$ are the energetic differences between the isotopologues of the reactant at the beginning of the reaction and $\Delta(\Delta G_{TS})$ are energetic differences between the isotopologues in the transition state (modified from Elsner et al., 2010).

assessable by comparing the isotope ratio of elements involved in the process (Aelion et al., 2010). The quantification of isotope fractionation is explained below.

1.7.2 Stable isotope analysis of carbon and natural attenuation

Carbon exists mainly in two stable isotopic forms, ^{12}C and ^{13}C, with a mean natural isotope ratio of $^{13}C/^{12}C$ = 0.001123, which means that the natural abundance of ^{13}C is 1.11 %. Application of stable carbon isotopes analysis has gained great attention in environmental studies since last decade. The most common technique for this purpose is isotope ratio mass spectrometry (IRMS), either coupled to a gas chromatograph (GC-IRMS) or to an elemental analyser (EA-IRMS) (Swartjes, 2011). The international standard used for reporting stable carbon isotope composition is Vienna Pee Dee Belemnite (VPDB) (eq. 1) (Coplen, 2011).

$$\delta^{13}C(‰) = \left(\frac{R_{sample}}{R_{reference}} - 1\right) \times 1000 \qquad (1)$$

The Rayleigh equation can be applied for mathematical description of microbial carbon stable isotope fractionation processes (eq. 2), where δ_t is the isotope ratio of the substrate at a certain time t of biodegradation, δ_0 is the initial isotope ratio of the substrate, C_{Bt}/C_0 is the fraction of substrate remaining during biodegradation at a certain time t, whereas C_{Bt} is the concentration of substrate at time t and C_0 is concentration at time 0, and ϵ is the enrichment factor.

$$\frac{(\delta_t+1)}{(\delta_0+1)} = \left(\frac{C_{Bt}}{C_0}\right)^\epsilon \quad (2)$$

The extent of contaminant biodegradation can be expressed as percentage of the initial contaminant concentration removed due to biodegradation (see eq. 3).

$$B[\%] = \left(1 - \frac{C_{Bt}}{C_0}\right) \cdot 100 \quad (3)$$

Combining this expression with the Rayleigh equation (eq. 2) allows the quantification of contaminant biodegradation within a time interval or along a flow path of the contaminant in a given environment using measured isotope ratios. Required data are the initial isotope ratio of the contaminant at a starting point in time or in space (generally the contaminant source) and the isotope ratio of the remaining contaminant at a temporal or spatial observation point (generally a well within the downstream of a source). The amount of contaminant degraded between the initial and a certain time t is then given by according to eq. 4 (Thullner et al., 2012).

$$B[\%] = \left(1 - \frac{C_{Bt}}{C_0}\right) \cdot 100 = \left[1 - \left(\frac{\delta_t+1}{\delta_0+1}\right)^{\left(\frac{1}{\epsilon}\right)}\right] \cdot 100 \quad (4)$$

1.8 Enantiomeric Fraction

The enantiomeric fraction (EF) is used to explain the relationship between enantiomers during biodegradation (Harner et al., 2000; Wiberg et al., 2001). The EF (+) is defined as $A^+/(A^+ + A^-)$, where A^+ and A^- correspond to the peak area or concentrations of (+) and (-) enantiomers (Harner et al., 2000). Racemic compounds have an EF (+) equal to 0.5. An EF (+) > 0.5 shows the preferential degradation of (-) enantiomer, and an EF (+) < 0.5 indicates the preferential degradation of (+) enantiomer. EF (-) is defined as $A^-/(A^+ + A^-)$.

1.8.1 Quantification of enantiomeric fractionation

Recently, Gasser et al. (2012) proposed to quantitatively describe the fractionation of enantiomers induced by biotransformation using the simplified version of the Rayleigh equation (eq. 2). Since, the assumption $R+1 \approx 1$ for the enantiomeric ratios which can be expressed by enantiomeric fractions (R = EF(-)/EF(+)) is not valid, the most general form of the Rayleigh equation (eq. 4) has to be used for the determination of enantiomeric enrichment factor (ε_e). First, eq. 4 is logarithmized and $\ln[C_t/C_0/\{(EF(-)t/EF(+)t+1)/(EF(-)0/EF(+)0+1)\}]$ is plotted versus $\ln[\{EF(-)t/EF(+)t)/(EF(-)0/EF(+)0\}]$, where C_t and C_0 are the sum of the concentration of both enantiomers at a given time (t) and prior to biodegradation (0), respectively. Then, the enantiomeric fractionation factor (ε_e) is obtained from the slope of the linear regression m = ε_e (see Chapter 3.1).

1.9   Application of CSIA for HCH transformations

To apply CSIA for the investigation of *in situ* HCH transformation, it is important to perform laboratory reference culture experiments to estimate carbon stable isotope enrichment factors (ε_c) and to characterize the degradation pathways by considering different environmental conditions. Preliminary laboratory investigations showed significant carbon stable isotope fractionation for γ-HCH during dechlorination under sulfate reducing conditions suggesting that CSIA could be applied to monitor γ-HCH degradation in anoxic environments (Badea et al., 2009). In a subsequent study, Badea and colleagues developed a method to analyse the changes in isotopic composition of individual enantiomers of α-HCH and demonstrated the applicability of the innovative enantiomer-specific stable isotope analysis (ESIA) concept for the anaerobic biotransformation with *Clostridium pasteurianum* (Badea et al., 2011). These studies provided the basis for the application of CSIA as tool to characterize and assess HCH transformation processes but there were still scientific gaps which needed to be addressed to prove its validation.

To fill these gaps of knowledge, therefore, the objectives of this book were:

- To establish carbon enrichment factors for HCH isomers and enantiomers for oxic environmental condition (Chapter 3.1).
- To validate enantiomer selective stable isotope analysis (ESIA) for assessing the fate of a chiral isomer (α-HCH) (Chapter 3.1).
- To compare carbon enrichment factors (ε_c) of pure cultures and mixed anaerobic cultures to validate representative enrichment factors for *in situ* microbial environment. (Chapter 3.2).
- To develop compound specific stable isotope analysis as a reference for identification of abiotic reactions (Chapter 3.3).
- To apply CSIA at real field conditions for assessing *in situ* transformation of HCH (Chapter 3.4).

References

Aelion C. M., Höhener, P., Hunkeler, D., & Aravena, R. (Eds.). (2009).Environmental isotopes in biodegradation and bioremediation, 1st edn. CRC Press, Boca Raton.

Anderson T. A., Guthrie E.A., Walton B.T. (1993) Bioremediation in the rhizosphere. Environmental Science & Technology, 27(13):2630-2636.

Bachmann A., De Bruin W., Jumelet J., Rijnaarts H., Zehnder A. (1988a) Aerobic biomineralization of α-hexachlorocyclohexane in contaminated soil. Applied and Environmental Microbiology, 54(2):548-554.

Bachmann A., Walet P., Wijnen P., De Bruin W., Huntjens J., Roelofsen W., Zehnder A. (1988b) Biodegradation of α-and β-hexachlorocyclohexane in a soil slurry under different redox conditions. Applied and Environmental Microbiology, 54(1):143-149.

Badea S.L., Vogt C., Weber S., Danet A.F., Richnow H. H. (2009) Stable isotope fractionation of γ-Hexachlorocyclohexane (Lindane) during reductive dechlorination by two strains of sulfate-reducing bacteria. Environmental Science & Technology, 43(9):3155-3161.

Badea S.L., Vogt C., Gehre M., Fischer A., Danet A. F., Richnow H. H. (2011) Development of an enantiomer-specific stable carbon isotope analysis (ESIA) method for assessing the fate of α-hexachlorocyclo-hexane in the environment. Rapid Communications in Mass Spectrometry, 25(10):1363-1372.

Baker K.H., Herson D.S. (1994) Bioremediation McGraw-Hill, New York.

Beurskens J.E., Stams A.J., Zehnder A.J., Bachmann A. (1991) Relative biochemical reactivity of three hexachlorocyclohexane isomers. Ecotoxicology and Environmental Safety, 21(2):128-136.

Bigeleisen J. (2004). "The relative reaction velocities of isotopic molecules." The Journal of Chemical Physics, 17:675-678.

Bintein S., Devillers J. (1996) Evaluating the environmental fate of lindane in France. Chemosphere, 32(12):2427-2440.

Bockelmann A., Zamfirescu D., Ptak T., Grathwohl P., Teutsch G. (2003) Quantification of mass fluxes and natural attenuation rates at an industrial site with a limited monitoring network: a case study. Journal of Contaminant Hydrology, 60(1):97-121.

Böltner D., Moreno-Morillas S., Ramos J.L. (2005) 16S rDNA phylogeny and distribution of lin genes in novel hexachlorocyclohexane-degrading Sphingomonas strains. Environmental Microbiology, 7(9):1329-1338.

Bombach P., Richnow H.H., Kastner M., Fischer A. (2010) Current approaches for the assessment of in situ biodegradation. Applied Microbiology and Biotechnology, 86(3):839-852.

Breivik K., Pacyna J.M., Münch J. (1999) Use of α-, β-and γ-hexachlorocyclohexane in Europe, 1970-1996. Science of the Total Environment, 239(1):151-163.

Buser H.-R., Mueller M.D. (1995) Isomer and enantioselective degradation of hexachlorocyclohexane isomers in sewage sludge under anaerobic conditions. Environmental Science & Technology, 29(3):664-672.

Coplen T.B. (2011) Guidelines and recommended terms for expression of stable-isotope-ratio and gas-ratio measurement results. Rapid Communication in Mass Spectrometry, 25(17):2538-2560.

Dadhwal M., Jit S., Kumari H., Lal R. (2009) *Sphingobium chinhatense* sp. nov., a hexachlorocyclohexane (HCH)-degrading bacterium isolated from an HCH dumpsite. International Journal of Systematic and Evolutionary Microbiology, 59(12):3140-3144.

De Paolis M., Lippi D., Guerriero E., Polcaro C., Donati E. (2013) Biodegradation of α-, β-, and γ-Hexachlorocyclohexane by *Arthrobacter fluorescens* and *Arthrobacter giacomelloi*. Applied Biochemistry and Biotechnology, 170:514-524

Deo P.G., Karanth N.G., Gopalakrishna N., Karanth K. (1994) Biodegradation of hexachlorocyclohexane isomers in soil and food environment. Critical Reviews in Microbiology, 20(1):57-78.

Doesburg W., Eekert M.H., Middeldorp P.J., Balk M., Schraa G., Stams A.J. (2005) Reductive dechlorination of β-hexachlorocyclohexane (β-HCH) by a Dehalobacter species in coculture with a Sedimentibacter sp. FEMS Microbiology Ecology, 54(1):87-95.

Dorsey A. (2005) Toxicological profile for alpha-, beta-, gamma-, and delta-Hexachlorocyclohexane US Department of Health & Human Services, Public Health Service, Agency for Toxic Substances and Disease Registry.

Elliott D.W., Lien H.-L., Zhang W.-X. (2009) Degradation of lindane by zero-valent iron nanoparticles. Journal of Environmental Engineering, 135(5):317-324.

Elsner M. (2010) Stable isotope fractionation to investigate natural transformation mechanisms of organic contaminants: principles, prospects and limitations. Journal of Environmental Monitoring, 12(11):2005-2031.

EPA (1999) Use of monitored natural attenuation at superfund, RCRA corrective action, and underground storage tank sites. Office of solid waste and emergency response, Washington DC

Feisthauer S., Seidel M., Bombach P., Traube S., Knöller K., Wange M., Fachmann S., Richnow H.H. (2012) Characterization of the relationship between microbial degradation processes at a hydrocarbon contaminated site using isotopic methods. Journal of Contaminant Hydrology, 133:17-29.

Feldmann R.J., Maibach H.I. (1974) Percutaneous penetration of some pesticides and herbicides in man. Toxicology and Applied Pharmacology, 28(1):126-132.

Gasser G., Pankratov I., Elhanany S., Werner P., Gun J., Gelman F., Lev O. (2012) Field and laboratory studies of the fate and enantiomeric enrichment of venlafaxine and O-desmethylvenlafaxine under aerobic and anaerobic conditions. Chemosphere, 88:98-105.

Geueke B., Garg N., Ghosh S., Fleischmann T., Holliger C., Lal R., Kohler H.P.E. (2013) Metabolomics of hexachlorocyclohexane (HCH) transformation: ratio of

LinA to LinB determines metabolic fate of HCH isomers. Environmental Microbiology, 15(4):1040-1049.

Geyer H.J., Scheunert I., Korte F. (1987) Correlation between the bioconcentration potential of organic environmental chemicals in humans and their n-octanol/water partition coefficients. Chemosphere, 16(1):239-252.

Hamada M., Kawano E., Kawamura S., Shiro M. (1982) A New isomer of 1,2,3,4,5-pentachlorocyclohexane from UV Irradiation products of α-isomers, β-isomers and δ-Isomers of 1,2,3,4,5,6-hexachlorocyclohexane. Agricultural and Biological Chemistry, 46(1):153-157.

Harner T., Jantunen L.M.M., Bidleman T.F., Barrie L.A., Kylin H., Strachan W.M.J., Macdonald R.W. (2000) Microbial degradation is a key elimination pathway of hexachlorocyclohexanes from the Arctic Ocean. Geophysical Research Letters, 27(8):1155-1158.

Hatzinger P.B., Böhlke J., Sturchio N.C. (2013) Application of stable isotope ratio analysis for biodegradation monitoring in groundwater. Current Opinion in Biotechnology, 24(3):542-549.

Haugen J.E., Wania F., Ritter N., Schlabach M. (1998) Hexachlorocyclohexanes in air in southern Norway. Temporal variation, source allocation, and temperature dependence. Environmental Science & Technology, 32(2):217-224.

Helm P.A., Diamond M.L., Semkin R., Bidleman T.F. (2000) Degradation as a loss mechanism in the fate of α-hexachlorocyclohexane in Arctic watersheds. Environmental Science & Technology, 34(5):812-818.

Heritage A., MacRae I. (1977) Degradation of lindane by cell-free preparations of *Clostridium sphenoides*. Applied and Environmental Microbiology, 34(2):222-224.

Hinckley D.A., Bidleman T.F., Rice C.P. (1991) Atmospheric organochlorine pollutants and air-sea exchange of hexachlorocyclohexane in the Bering and Chukchi seas. Journal of Geophysical Research: Oceans, (1978–2012) 96(C4):7201-7213.

Hofstetter T.B., Berg M. (2011) Assessing transformation processes of organic contaminants by compound-specific stable isotope analysis. TrAC Trends in Analytical Chemistry, 30(4):618-627.

Imai R., Nagata Y., Fukuda M., Takagi M., Yano K. (1991) Molecular cloning of a Pseudomonas paucimobilis gene encoding a 17-kilodalton polypeptide that eliminates HCl molecules from γ-hexachlorocyclohexane. Journal of Bacteriology, 173(21):6811-6819.

Ito M., Prokop, Z., Klvaňa, M., Otsubo, Y., Tsuda, M., Damborský, J., & Nagata, Y. (2007). Degradation of β-hexachlorocyclohexane by haloalkane dehalogenase LinB from γ-hexachlorocyclohexane-utilizing bacterium *Sphingobium* sp. MI1205. Archives of Microbiology, 188(4):313-325.

Johri A.K., Dua M., Tuteja D., Saxena R., Saxena D., Lal R. (1996) Genetic manipulations of microorganisms for the degradation of hexachlorocyclohexane. FEMS Microbiology Reviews, 19(2):69-84.

Jones A., Panagos P., Barcelo S., Bouraoui F., Bosco C., Dewitte O., Gardi C., Hervás J., Hiederer R., Jeffery S. (2012) The State of Soil in Europe-A contribution of the jrc to the european environment agency's environment state and Outlook Report-SOER 2010.

Jung D., Becher H., Edler L., Flesch-Janys D., Gurn P., Konietzko J., Manz A., Papke O. (1997) Elimination of β-hexachlorocyclohexane in occupationally exposed persons. Journal of Toxicology and Environmental Health, 51(1):23-34.

Jury W., Spencer W., Farmer W. (1983) Behavior assessment model for trace organics in soil: I. Model description. Journal of Environmental Quality, 12(4):558-564.

Lal R., Pandey G., Sharma P., Kumari K., Malhotra S., Pandey R., Raina V., Kohler H.-P.E., Holliger C., Jackson C. (2010) Biochemistry of microbial degradation of hexachlorocyclohexane and prospects for bioremediation. Microbiology and Molecular Biology Reviews, 74(1):58-80.

Langenhoff A.A., Staps S.J., Pijls C., Rijnaarts H.H. (2013) Stimulation of hexachlorocyclohexane (HCH) biodegradation in a full scale *in situ* bioscreen. Environmental Science & Technology, 47(19):11182-11188.

Law S.A., Bidleman T.F., Martin M.J., Ruby M.V. (2004) Evidence of enantioselective degradation of α-hexachlorocyclohexane in groundwater. Environmental Science & Technology, 38(6):1633-1638.

Li Y.F. (1999) Global technical hexachlorocyclohexane usage and its contamination consequences in the environment: from 1948 to 1997. Science of The Total Environment, 232(3):121-158.

Lichtenstein E., Polivka J. (1959) Persistence of some chlorinated hydrocarbon insecticides in turf Soils 1. Journal of Economic Entomology, 52(2):289-293.

Lide D.R. (1996). CRC handbook of chemistry and physics: a ready-reference book of chemical and physical data. 76th. ed.

Ma A.Z., Wu J., Zhang G.-s., Wang T., Li S. (2005) Isolation and characterization of a HCH degradation Sphingomanas sp. stain BHC-A]. Wei sheng Wu Xue Bao-Acta Microbiologica Sinica, 45(5):728-732.

MacRae I.C., Raghu K., Castro T.F. (1967) Persistence and biodegradation of four common isomers of benzene hexachloride in submerged soils. Journal of Agricultural and Food Chemistry, 15(5):911-914.

Macrae I.C., Raghu K., Bautista E.M. (1969) Anaerobic degradation of insecticide Lindane by *Clostridium* Sp. Nature, 221(5183):859-860.

Manickam N., Mau M., Schlömann M. (2006) Characterization of the novel HCH-degrading strain, *Microbacterium* sp. ITRC1. Applied Microbiology and Biotechnology, 69(5):580-588.

Martin H., Patterson B.M., Davis G.B., Grathwohl P. (2003) Field trial of contaminant groundwater monitoring: comparing time-integrating ceramic dosimeters and conventional water sampling. Environmental Science & Technology, 37(7):1360-1364.

Mehboob F., Langenhoff, A.A., Schraa, G., & Stams, A.J. (2013). Anaerobic degradation of lindane and other HCH Isomers. In Management of Microbial Resources in the Environment (pp. 495-521). Springer Netherlands.

Mohn W.W., Garmendia J., Galvao T.C., De Lorenzo V. (2006) Surveying biotransformations with à la carte genetic traps: translating dehydrochlorination of lindane (γ-hexachlorocyclohexane) into lacZ-based phenotypes. Environmental Microbiology, 8(3):546-555.

Nagasawa S., Kikuchi R., Nagata Y., Takagi M., Matsuo M. (1993) Aerobic mineralization of γ-HCH by *Pseudomonas paucimobilis* UT26. Chemosphere, 26(9):1719-1728.

Nagata Y., Futamura A., Miyauchi K., Takagi M. (1999) Two different types of dehalogenases, LinA and LinB, involved in γ-Hexachlorocyclohexane degradation in *Sphingomonas paucimobilis* UT26 are localized in the periplasmic space without molecular processing. Journal of Bacteriology, 181(17):5409-5413.

Ngabe B., Bidleman T.F., Falconer R.L. (1993) Base hydrolysis of α-hexachlorocyclohexanes and γ-hexachlorocyclohexanes. Environmental Science & Technology, 27(9):1930-1933.

Norris R. D. (1995). In-situ bioremediation of ground water and geological material: A review of technologies. DIANE Publishing

Olivero-Verbel, J., Guerrero-Castilla, A., & Ramos, N. R. (2011). Biochemical effects induced by the hexachlorocyclohexanes. In Reviews of Environmental Contamination and Toxicology Volume 212 (pp. 1-28). Springer New York.

Ørbæk J.B., Tombre I., Kallenborn R. (2004) Challenges in Arctic-Alpine environmental research. Arctic, Antarctic, and Alpine Research, 3(3):281-283.

Österreicher-Cunha P., Langenbach T., Torres J.P., Lima A.L., de Campos T.M., Vargas Jr E.p.d.A., Wagener A.R. (2003) HCH distribution and microbial parameters after liming of a heavily contaminated soil in Rio de Janeiro. Environmental Research, 93(3):316-327.

Pal R., Bala S., Dadhwal M., Kumar M., Dhingra G., Prakash O., Prabagaran S., Shivaji S., Cullum J., Holliger C. (2005) Hexachlorocyclohexane-degrading bacterial strains *Sphingomonas paucimobilis* B90A, UT26 and Sp+, having similar lin genes, represent three distinct species, *Sphingobium indicum* sp. nov., *Sphingobium japonicum* sp. nov. and *Sphingobium francense* sp. nov., and reclassification of [Sphingomonas] chungbukensis as *Sphingobium chungbukense* comb. nov. International Journal of Systematic and Evolutionary Microbiology, 55(5):1965-1972.

Paraiba L.C., Spadotto C.A. (2002) Soil temperature effect in calculating attenuation and retardation factors. Chemosphere, 48(9):905-912.

Pardieck D.L., Bouwer E.J., Stone A.T. (1992) Hydrogen peroxide use to increase oxidant capacity for in situ bioremediation of contaminated soils and aquifers: A review. Journal of Contaminant Hydrology, 9(3):221-242.

Phillips T.M., Seech A.G., Lee H., Trevors J.T. (2005) Biodegradation of hexachlorocyclohexane (HCH) by microorganisms. Biodegradation, 16(4):363-392.

Pope D.F., Acree S.D., Levine H., Mangion S., Van Ee J., Hurt K., Wilson B., Burden D.S. (2004) Performance monitoring of MNA remedies for VOCs in ground water US Environmental Protection Agency, National Risk Management Research Laboratory.

Ricking M., Schwarzbauer J. (2008) HCH residues in point-source contaminated samples of the Teltow Canal in Berlin, Germany. Environmental Chemistry Letters, 6(2):83-89.

Sagelsdorff P., Lutz W.K., Schlatter C. (1983) The relevance of covalent binding to mouse liver DNA to the carcinogenic action of hexachlorocyclohexane isomers. Carcinogenesis, 4(10):1267-1273.

Sahu S.K., Patnaik K.K., Sharmila M., Sethunathan N. (1990) Degradation of α-, β-, and γ-hexachlorocyclohexane by a soil bacterium under aerobic conditions. Applied and Environmental Microbiology, 56(11):3620-3622.

Sahu S.K., Patnaik K., Bhuyan S., Sethunathan N. (1993) Degradation of soil-applied isomers of hexachlorocyclohexane by a *Pseudomonas* sp. Soil Biology and Biochemistry, 25(3):387-391.

Sahu S.K., Patnaik K., Bhuyan S., Sreedharan B., Kurihara N., Adhya T., Sethunathan N. (1995) Mineralization of. alpha.-,. gamma.-, and. beta.- isomers of Hexachlorocyclohexane by a soil bacterium under aerobic conditions. Journal of Agricultural and Food Chemistry, 43(3):833-837.

Schirmer M., Dahmke A., Dietrich P., Dietze M., Gödeke S., Richnow H.H., Schirmer K., Weiß H., Teutsch G. (2006) Natural attenuation research at the contaminated megasite Zeitz. Journal of Hydrology, 328(3):393-407.

Senoo K., Wada H. (1989) Isolation and identification of an aerobic γ-HCH- decomposing bacterium from soil. Soil Science and Plant Nutrition, 35(1):79-87.

Sharma H.D., & Reddy, K. R. (2004). Geoenvironmental engineering: site remediation, waste containment, and emerging waste management technologies. John Wiley & Sons, Inc.

Singh R., Misra V., Singh R.P. (2011) Remediation of-hexachlorocyclohexane contaminated soil using nanoscale zero-valent iron. Journal of Bionanoscience, 5(1):82-87.

Slade R. (1945) The γ isomer of Hexachlorocyclohexane ('Gammexane'): An insecticide with outstanding properties - The hurter memorial lecture to liverpool section, Chemistry and Industry, 40:314

Steinmetz R., Young P.C., Caperell-Grant A., Gize E.A., Madhukar B.V., Ben-Jonathan N., Bigsby R.M. (1996) Novel estrogenic action of the pesticide residue β-hexachlorocyclohexane in human breast cancer cells. Cancer Research, 56(23):5403-5409.

Stewart D., Chisholm D. (1971) Long-term persistence of BHC, DDT and Chlordane in a sandy loam soil. Canadian Journal of Soil Science, 51(3):379-383.

Suar M., Hauser A., Poiger T., Buser H.R., Müller M.D., Dogra C., Raina V., Holliger C., van der Meer J.R., Lal R. (2005) Enantioselective transformation of α-hexachlorocyclohexane by the dehydrochlorinases LinA1 and LinA2 from the soil bacterium *Sphingomonas paucimobilis* B90A. Applied and Environmental Microbiology, 71(12):8514-8518.

Swartjes F. A. (2011). Dealing with contaminated sites (pp. 1079-1104). Netherlands: Springer.

Thomas J.C., Berger F., Jacquier M., Bernillon D., Baud-Grasset F., Truffaut N., Normand P., Vogel T.M., Simonet P. (1996) Isolation and characterization of a novel γ-hexachlorocyclohexane-degrading bacterium. Journal of Bacteriology, 178(20):6049-6055.

Thullner M., Centler F., Richnow H.H., Fischer A. (2012) Quantification of organic pollutant degradation in contaminated aquifers using compound-specific stable isotope analysis – review of recent developments. Organic Geochemistry, 42(12):1440-1460.

Van Liere H., Staps S., Pijls C., Zwiep G., Lassche R., Langenhoff A. (2003) Full scale case: successful in situ bioremediation of a HCH contaminated industrial site in central Europe (The Netherlands), Forum book :128-132.

Vijgen J. (2006). The legacy of lindane HCH isomer production. Main Report, IHPA, Holte, January.

Vijgen J., Abhilash P., Li Y.F., Lal R., Forter M., Torres J., Singh N., Yunus M., Tian C., Schäffer A. (2011) Hexachlorocyclohexane (HCH) as new Stockholm Convention POPs-a global perspective on the management of Lindane and its waste isomers. Environmental Science and Pollution Research, 18(2):152-162.

Voldner E.C., Li Y.-F. (1995) Global usage of selected persistent organochlorines. Science of the Total Environment, 160:201-210.

Weber R., Gaus C., Tysklind M., Johnston P., Forter M., Hollert H., Heinisch E., Holoubek I., Lloyd-Smith M., Masunaga S. (2008) Dioxin-and POP-contaminated sites-contemporary and future relevance and challenges. Environmental Science and Pollution Research, 15(5):363-393.

Wiberg K., Brorstrom-Lunden E., Wangberg I., Bidleman T.F., Haglund P. (2001) Concentrations and fluxes of hexachlorocyclohexanes and chiral composition of α-HCH in environmental samples from the southern Baltic Sea. Environmental Science & Technology, 35:4739-4746.

Wiedemeier T. H. (Ed.). (1999). Natural attenuation of fuels and chlorinated solvents in the subsurface. John Wiley & Sons.

Wiegert C. (2013) Application of two dimensional compound specific carbon-chlorine isotope analyses for degradation monitoring and assessment of organic pollutants in contaminated soil and groundwater, Department of Applied Environmental Science (ITM), Stockholm University, Stockholm:1-52.

Wijker R.S., Bolotin J., Nishino S.F., Spain J.C., Hofstetter T.B. (2013) Using compound-specific isotope analysis to assess biodegradation of nitroaromatic explosives in the subsurface. Environmental Science & Technology, 47(13): 6872-6883

Willett K.L., Ulrich E.M., Hites R.A. (1998) Differential toxicity and environmental fates of hexachlorocyclohexane isomers. Environmental Science & Technology, 32(15):2197-2207.

Wycisk P., Stollberg R., Neumann C., Gossel W., Weiss H., Weber R. (2013) Integrated methodology for assessing the HCH groundwater pollution at the multi-source contaminated mega-site Bitterfeld/Wolfen. Environmental Science and Pollution Research, 20(4):1907-1917.

Zhang W.X. (2003) Nanoscale iron particles for environmental remediation: an overview. Journal of Nanoparticle Research, 5(3-4):323-332.

Objectives & structure of the book

2 Objective and structure of book

To fill the above mentioned scientific gaps for the application of CSIA as a tool to assess *in situ* transformation of HCH by various laboratory and field investigations. The overall objective of this book, therefore, was to validate the application of carbon stable isotope analysis (CSIA) and enantiomer specific stable isotope analysis (ESIA) to characterize biotic and abiotic transformation of HCH *in situ*. Additionally, to prove that CSIA in combination with enantiomer-selective degradation of α-HCH can be applied as an effective and reliable tool for monitoring natural attenuation of HCH.

The specific objectives of the book were:

- Evaluation of CSIA for characterization of HCH biotransformation.
- Evaluation of enantiomer fractionation (EF) and enantiomer-specific stable isotope analysis (ESIA) as an additional technique in addition to CSIA for assessing the fate of a chiral isomer (α-HCH).
- Evaluation of anaerobic HCH biotransformation and pathways by combining metabolite analysis and CSIA.
- Characterisation of isotope fractionation of HCH during environmentally relevant abiotic reactions.
- Application of CSIA concepts *in situ* to assess biotransformation and contaminant source identification.

This doctoral book is written in chapter's format and is based on following four main results chapters.

Chapter 3.1 focuses on the carbon stable isotope fractionation during aerobic degradation of α and γ-HCH and to refine the concept of ESIA. The carbon stable isotope fractionation during aerobic degradation of the enantiomers of α-HCH by two *Sphingobium* spp. *S. indicum* strain B90A and *S. japonicum* strain UT26 were determined, and compared to the isotope fractionation of bulk α-HCH and γ-HCH. We further investigated changes in enantiomeric composition of α-HCH during aerobic biodegradation in order to verify if the Rayleigh equation can be applied for describing the fractionation of the α-HCH enantiomers. We therefore propose that

ESIA, in addition to CSIA, can be used as a tool for analyzing the fate of enantiomeric chemicals in the environment making use of both enantioselective degradation and isotope fractionation as indicators.

Chapter 3.2 focuses on the biotransformation of HCHs by *Dehalococcoides mccartyi* strain 195 and strain BTF08 and related growth. The degradation pathway was investigated by analysing the resulting metabolites and end products. Shifts in stable carbon isotope composition of γ-HCH were compared to the co-metabolic HCH degrading *Clostridium pasterianum* as well as to a γ-HCH degrading enrichment culture.

Chapter 3.3 investigates carbon stable isotope fractionation during environmentally relevant abiotic transformation reactions of α-HCH which include photo-induced reactions, hydrolysis, electrochemical reaction and reductive reaction by zerovalent iron nanoparticles (nZVI).

Chapter 3.4 demonstrates the application of CSIA to assess *in situ* biotransformation of HCH by using carbon isotope ratios of HCHs in combination with pollutant concentration patterns and hydrogeochemical information from the field site. Additionally, this study deals with the application of CSIA as tool to identify contaminant sources. Overall, the applicability of CSIA was validated as an appropriate alternate monitoring tool for the implementation and successful control of innovative management and remediation concepts like Monitored or Enhanced Natural Attenuation (MNA, ENA).

3

Results

3.1

Enantioselective Carbon Stable Isotope Fractionation of Hexachlorocyclohexane during Aerobic Biodegradation by *Sphingobium* spp.

Safdar Bashir,[†] Anko Fischer,[†,‡] Ivonne Nijenhuis,[†] and Hans-Hermann Richnow[†]

[†]Department of Isotope Biogeochemistry, Helmholtz Centre for Environmental Research – UFZ, Permoserstraße 15, 04318, Leipzig,
Germany

[‡]Isodetect - Company for Isotope Monitoring, Permoserstraße 15, 04318 Leipzig, Germany

Abstract

Carbon isotope fractionation was investigated for the biotransformation of γ- and α-Hexachlorocyclohexane as well as enantiomers of α-HCH using two aerobic bacterial strains: *Sphingobium indicum* strain B90A and *Sphingobium japonicum* strain UT26. Carbon isotope enrichment factors (ε_c) for γ-HCH (ε_c = -1.5 ± 0.1 ‰ and -1.7 ± 0.2 ‰) and α-HCH (ε_c = -1.0 ± 0.2 ‰ and -1.6 ± 0.3 ‰) were similar for both aerobic strains, but lower in comparison with previously reported values for anaerobic γ-HCH degradation. Isotope fractionation of α-HCH enantiomers was higher (+) α-HCH (ε_c = -2.4 ± 0.8 ‰ and -3.3 ± 0.8 ‰) in comparison to (-) α-HCH (ε_c = -0.7 ± 0.2 ‰ and -1.0 ± 0.6 ‰). The microbial fractionation between the α-HCH enantiomers was quantified by the Rayleigh equation and enantiomeric fractionation factors (ε_e) for *S. indicum* strain B90A and *S. japonicum* strain UT26 were -42.8 ± 15.5 % and -22 ± 5.7 %, respectively. The extent and range of isomer and enantiomeric carbon isotope fractionation of HCHs with *Sphingobium* spp. suggests that aerobic biodegradation of HCHs can be monitored *in situ* by carbon stable isotope analysis (CSIA) and enantiomer-specific isotope analysis (ESIA). In addition, enantiomeric fractionation has the potential as a complementary approach to CSIA and ESIA for assessing the biodegradation of α-HCH at contaminated field sites.

3.1.1 Introduction

About 25% of worldwide applied organic chemicals, e.g. pharmaceuticals or pesticides, are chiral and were applied as mixtures of isomers and/or enantiomers (Hegeman and Laane, 2001). Despite the almost similar molecular structure and identical physical properties, however, enantiomers may have different characteristics related to (biogeo)chemical reactions governing their persistence and toxicity in the environment (Garrison, 2006; Kallenborn, 2001; Kohler et al., 1997). This motivates our investigation for studying the behavior of enantiomers combined with stable isotope techniques to gain further information for tracing their fate in the environment. For this purpose, we selected the hexachlorocyclohexane (HCH) of which α-HCH is chiral.

HCHs, comprising mainly α, β, γ and δ-HCH, were among the most produced and applied insecticides, as technical HCH or Lindane (γ-HCH), between 1950 and 2000 (Breivik et al., 1999; Li, 1999; Li and Macdonald, 2005). Technical HCH, containing 60-70 % of the chiral isomer α-HCH, 5 to 12 % of β-HCH, 10 to 12 % of γ-HCH, 6 to 10 % δ-HCH and 3 to 4 % ε- HCH of which only γ-HCH has specific insecticidal activity, was extensively used mainly in developing countries (Bhatt et al., 2009; Li, 1999). In addition to technical HCH, the γ-HCH isomer known as Lindane was intensively used as a pure component in insecticide formulations (Voldner and Li, 1995). The large amount of by-products produced during the Lindane production (one ton of purified γ-HCH results in eight to twelve tons of HCH by-products containing all other isomers) were often disposed improperly resulting in point source contamination of soil and groundwater at recent and former HCH production sites (Vijgen et al., 2011) . Between four to seven million tons of technical HCH waste was estimated to be produced around the globe during 60 years of Lindane synbook. These residual of HCHs are toxic, persistent, and potentially bio-accumulative (Butte et al., 1991; Canton et al., 1975; Solomon et al., 1977; Willett et al., 1998) resulting in the recent inclusion as new persistent organic pollutants (POPs) in the Stockholm convention (Vijgen et al., 2011) . Therefore, concepts to trace the fate of HCH in the environment are essential.

Biodegradation is a major process removing HCHs in soil, aquifers and surface water bodies (Lal et al., 2010). Anaerobic transformation of HCH results in monochlorobenzene and benzene as accumulating metabolites via reductive beta-elimination (Quintero et al., 2006) while aerobic degradation proceeds via dehydrochlorination. The aerobic transformation of γ-HCH is characterized by two initial dehydrochlorination reactions producing the putative product 1,3,4,6-tetrachloro-1,4-cyclohexadiene (1,3,4,6-TCDN), via the intermediate γ-pentachlorocyclohexene (γ-PCCH) (Imai et al., 1991; Nagasawa et al., 1993a; Nagasawa et al., 1993b) whereby the other side end products 1,2,4-trichlorobenzene (1,2,4-TCB) and 2,5-dichloropenol (2,5-DCP) may be formed depending on the different reaction routes (Geueke et al., 2012). α-HCH biotransformation via dehydrochlorination proceeds similarly, resulting in β-PCCH, as the respective enantiomer from (+) and (-)-α-HCH, as the first intermediate which are then converted to 1,2,4-TCB (Suar et al., 2005) . In these reactions, LinA has been described as responsible enzyme for biotransformation of α-HCH and γ-HCH (Lal et al., 2010).

carbon stable isotope analysis (CSIA) has become a well-developed concept to determine *in situ* biodegradation of common contaminants (Swartjes, 2011). The stable isotope fractionation upon the bond cleavage of the first irreversible reaction leads to an enrichment of heavier isotopologues in the residual (non-degraded) pollutant fraction. The extent of stable isotope fractionation allows to assess pollutant biodegradation qualitatively and quantitatively as well as to elucidate reaction mechanisms (Hofstetter and Berg, 2011; Thullner et al., 2011). The compound-specific isotope enrichment factor (ε_c), correlating change in concentration to isotope enrichment, from laboratory reference studies is needed as a reference for application of CSIA concepts in field studies.

Significant carbon stable isotope fractionation for Lindane (γ-HCH) could be observed during dechlorination under sulfate reducing conditions suggesting that CSIA could be applied to monitor γ-HCH degradation in anoxic environments (Badea et al., 2009). In a subsequent study, Badea *et al.* developed a method to analyze the changes in isotopic composition of individual enantiomers of α-HCH and demonstrated the applicability of the innovative enantiomer-specific stable isotope analysis (ESIA) concept for the anaerobic biotransformation with *Clostridium pasteurianum* (Badea et al., 2011).

To refine the concept of ESIA, we determined the carbon stable isotope fractionation during aerobic degradation of the enantiomers of α-HCH by two *Sphingobium* spp. strains, *Sphingobium indicum* strain B90A and *Sphingobium japonicum* strain UT26, and compared this to the isotope fractionation of bulk α-HCH and γ-HCH. We further investigated changes in enantiomeric composition of α-HCH during aerobic biodegradation in order to verify if the Rayleigh equation can be applied for describing the fractionation of the α-HCH enantiomers. We therefore propose that ESIA, in addition to CSIA, can be used as a tool for analyzing the fate of enantiomeric chemicals in the environment making use of both enantioselective degradation and isotope fractionation as indicators.

3.1.2 Material and methods

Chemicals

γ-HCH (analytical purity, 97 %), α-HCH (99 %), δ-HCH (99.5 %), 1,2,4-TCB and toluene were purchased from Sigma Aldrich (Germany). Pentane (analytical purity > 99 %) was obtained from Carl Roth, Germany. β- and γ-pentachlorocyclohexene (PCCH) were produced by alkaline dehydrochlorination of α and γ-HCH as described previously (Trantírek et al., 2001).

Bacterial strains and culture conditions

S. *indicum* strain B90A was provided by Prof. Rup Lal, University of Delhi, India and was grown on LB agar as described previously at 28°C (Imai et al., 1989). S. *japonicum* strain UT26 was obtained from Prof. Jiri Damborsky, Masaryk University, Czech Republic and grown at 1/3 LB agar at 30°C (Nagata et al., 1999).

Biodegradation experiments

Biodegradation of α- and γ-HCH was studied in batch culture experiments at the previously described optimal growth conditions indicated below. Glass serum bottles (240 mL) were equipped with oxygen sensor spots to monitor the oxygen concentration as described previously (Rosell et al., 2009). Mineral medium (Kumari et al., 2002) (50 mL) (with 0.1 % glucose) (Raina et al., 2008) was added in the serum bottles and the respective HCH isomer was added at a final concentration of 18 µM from 0.1 M stock solutions in acetone to each bottle. Bacterial strains grown in LB medium up to an OD_{600}=0.5 (10^8 cells mL^{-1}) were used for inoculation (0.5 ml or 10^6 cells mL^{-1}). Bottles were sealed with Teflon® coated rubber stoppers and crimped. All batches were incubated at 30°C at 150 rpm. Abiotic controls without inoculum were incubated under identical conditions.

Sampling and extraction procedure

A sacrificial approach was adopted for sampling as described previously (Cichocka et al., 2008). Each bottle was sacrificed with two ml of a saturated Na_2SO_4 (280 g L^{-1}) solution which was adjusted to pH 1 with H_2SO_4 to stop biological activity. All sacrificed batches were stored at 4°C until extraction. Before extraction, two internal standards, which were used for the quantification of HCH and products as well as internal standard for isotope analysis, were added in the sacrificed bottles: 10 µM δ-HCH for the HCH isomers and 10 µM toluene for metabolites. The batches were placed on a rotary shaker for 1 hour at 120 rpm speed at 20°C and then 1 ml of *n*-pentane was added in each bottle. All bottles were placed on a shaker again at 4°C at 120 rpm for 2 hours. Then the *n*-pentane phase was removed using glass Pasteur pipettes. Extracts were stored at -20°C until analysis. The extraction efficiency was >95%. In addition, the extraction procedure did not lead to significant changes of carbon isotope ratios (data not shown).

Analytical methods

GC-MS: A gas chromatograph (GC), (7890A, Agilent Technologies, Palo, USA) coupled to a mass spectrometer (MS) (5975C, Agilent Technologies, Palo, USA) was used for identification and structural characterization of HCHs and their metabolites.

HCH isomers and their metabolites were separated by BPX-5 capillary column (30m x 0.25mm x 0.25 µM) (SGE, Darmstadt, Germany). The temperature program used for separation of HCH isomers and their metabolites was described elsewhere (Badea et al., 2009). For separation of enantiomers a γ-DEX™ 120 chiral column (column length x I.D.30 m x 0.25 mm, d_f =0.25 µM) (Supelco Bellefonte, Pennsylvania, USA) was used and the method was described previously (Badea et al., 2011).

GC-C-IRMS: The stable carbon isotope ratios of HCHs and their metabolites were analyzed by a gas chromatograph-combustion-isotope ratio mass spectrometer (GC-C-IRMS). The system consisted of GC (6890, Agilent Technologies, Palo, USA) coupled with Conflow III interface (Thermo Fisher Scientific, Bremen, Germany) to a MAT252 IRMS (Thermo Fisher Scientific, Bremen, Germany) as described previously (Badea et al., 2009). Three µL aliquots of *n*-pentane extract were injected with a split of 1:3 while the samples with concentration below 1 µM were run with splitless injection. All samples were measured in at least three replicates and the typical uncertainty of analysis was <0.5 ‰. The carbon isotope ratios were expressed in the delta notation ($δ^{13}C$) relative to the international standard Vienna Pee Dee Belemnite (V-PDB) according to eq. 1 (Coplen, 2011).

$$\delta^{13}C_{sample} = \frac{R_{sample}}{R_{standard}} - 1 \qquad (1)$$

R_{sample} and $R_{standard}$ were the $^{13}C/^{12}C$ ratios of the sample and the standard, respectively and the $δ^{13}C$-values were reported in part per thousand (‰).

Quantification of isotope fractionation

For the description of stable isotope fractionation of biochemical reactions the Rayleigh equation can be applied in its most general form (Mariotti et al., 1981)

$$\frac{R_t}{R_0} = \left(\frac{\frac{C_t}{C_0}}{\frac{R_t+1}{R_0+1}} \right)^{\varepsilon} \qquad (2)$$

where R_t, R_0 and C_t, C_0 are the stable isotope ratios (e.g., $^{13}C/^{12}C$) and concentrations of a compound at a given point in time (t) and at the beginning of a transformation reaction (0), respectively. The isotope enrichment factor ε correlates the changes in stable isotope ratios (R_t/R_0) with the changes in the concentrations (C_t/C_0).

For stabile carbon isotope ratios ($R = {}^{13}C/{}^{12}C$) with natural abundance, the assumption $R+1 \approx 1$ is valid and the simplified version of the Rayleigh equation can be used for the assessment of stable carbon isotope fractionation of biodegradation processes:

$$\frac{R_t}{R_0} = \left(\frac{C_t}{C_0}\right)^\varepsilon . \qquad (3)$$

The carbon isotope enrichment factor (ε_c) was determined from the logarithmic form of the Rayleigh equation,

$$\ln\left(\frac{(\delta_t{}^{13}C+1)}{(\delta_0{}^{13}C+1)}\right) = \varepsilon_c \ln\left(\frac{C_t}{C_0}\right) \qquad (4)$$

plotting $\ln(C_t/C_0)$ versus $\ln[(\delta_t{}^{13}C+1)/(\delta_0{}^{13}C+1)]$ and obtaining ε_c from the slope of the linear regression ($m = \varepsilon_c$). Since carbon isotope enrichment factors are typically small, ε_c-values were reported in part per thousand (‰). The error of the isotope enrichment factors is reported as 95% confidence interval (CI) determined by a regression curve analysis (Elsner et al., 2007).

The average carbon isotope composition of substrate and products ($\delta^{13}C_{avg}$) was calculated in a mass balance approach by multiplying the concentration of substrate or product (C_i) with its respective carbon isotope ratio ($\delta^{13}C_i$) divided by the sum concentration of substrate and products (C_{sum}).

$$\delta^{13}C_{avg}(‰) = \frac{\sum (C_i * \delta^{13}C_i)}{C_{sum}} \qquad (5)$$

Enantiomeric Fraction

The enantiomeric fraction (EF) is used to explain the relationship between enantiomers during biodegradation (Harner et al., 2000a; Wiberg et al., 2001). The EF (+) is defined as $A^+/(A^+ + A^-)$, where A^+ and A^- correspond to the peak area or concentrations of (+) and (-) enantiomers (Harner et al., 2000a). Racemic compounds have an EF (+) equal to 0.5. An EF (+) > 0.5 shows the preferential degradation of (-) enantiomer, and an EF (+) <0.5 indicates the preferential degradation of (+) enantiomer. EF (-) is defined as $A_-/(A_++A_-)$.

Quantification of enantiomeric fractionation

Recently, Gasser et al. (2012) proposed to quantitatively describe the fractionation of enantiomers induced by biotransformation using the simplified version of the Rayleigh equation (eq. 3) (Gasser et al., 2012). Since the assumption $R+1 \approx 1$ for the enantiomeric ratios which can be expressed by enantiomeric fractions (R = EF(-)/EF(+)) is not valid, the most general form of the Rayleigh equation (eq. 1) has to be used for the determination of enantiomeric enrichment factor (ε_e). First, eq. 1 is logarithmized and $\ln[C_t/C_0/\{(EF(-)_t/EF(+)_t+1)/(EF(-)_0/EF(+)_0+1)\}]$ is plotted versus $\ln[\{EF(-)_t/EF(+)_t\}/(EF(-)_0/EF(+)_0)]$, where C_t and C_0 are the sum of the concentration of both enantiomers at a given time (t) and prior to biodegradation (0), respectively. Then, the enantiomeric fractionation factor (ε_e) is obtained from the slope of the linear regression $m = \varepsilon_e$. The error of the enantiomeric enrichment factors is given as 95% confidence interval (CI) determined by a regression curve analysis.

3.1.3 Results and discussion

Biodegradation and carbon isotope fractionation of γ-HCH

Degradation experiments with *S. indicum* strain B90A and *S. japonicum* strain UT26 as inoculum were performed to assess the time course of γ-HCH degradation and the change in carbon isotope ratio of γ-HCH and its products. Both bacterial strains were able to degrade γ-HCH within 24 hours of incubation. However, the biodegradation of γ-HCH was faster by *S. indicum* strain B90A (3.9 µM h^{-1}) as compared to *S. japonicum* strain UT26 (1.2 µM h^{-1}). γ-HCH concentrations were stable in abiotic controls indicating that abiotic losses could be neglected (not shown). γ-pentachlorocyclohexene (γ-PCCH) was the first observed intermediate produced by both bacterial strains. Other metabolites observed were 1,2,4-TCB and 2,5-dichlorophenol (2,5-DCP). Complete removal of γ-HCH was observed with *S.*

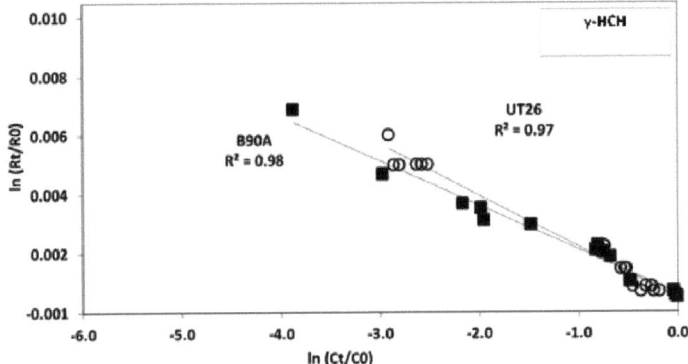

Figure 1: Linearized Rayleigh equation plots showing the carbon isotope fractionation for the biodegradation of γ-HCH by S. indicum strain B90A (closed square) and S. japonicum strain UT26 (open circle) and the correlation factors (R^2) of the linear regressions.

indicum strain B90A and complete removal of substrate was indicated by the mass and isotope balance calculated (Supporting information S1 A and B) (Appendix A1). In case of batches inoculated with S. japonicum strain UT26, the same metabolites were observed during biodegradation, however, γ-HCH was not completely degraded and 1,2,4-TCB and 2,5-dichlorophenol (2,5-DCP) were end-products as confirmed by the mass and isotope balance (Supporting information, Figures S1 C and D, supporting information, Appendix A1). Biodegradation stopped in case of S. japonicum strain UT26, when approx. one µM of γ-HCH was remained, which may be due to the toxic effects of metabolites as described previously (Endo et al., 2006).

The carbon isotope ratio of γ-HCH at the beginning of the experiment was -27.6 ± 0.5 ‰. In abiotic controls, this value remained constant during whole experiment (data not shown) confirming that besides biodegradation no other processes lead to changes in carbon isotope ratios of γ-HCH. In live culture batches, $\delta^{13}C$ values of substrate enriched significantly during the degradation experiment. In batches inoculated with S. indicum strain B90A, the $\delta^{13}C$ composition shifted from -27.5 ± 0.1 ‰ to -19.5 ± 0.3 ‰ upon 97 % removal. The mass balance approach calculating the average concentration and isotope composition of substrate and products showed a mass removal correlated with an ^{13}C-enrichment in cumulative stable isotope composition indicating the degradation of HCH and its detectable metabolites for *S. indicum* strain B90A (Supporting information, Figure S1 B) (Appendix A1). In case of *S. japonicum* strain UT26, 94 % biodegradation of γ-HCH resulted in changes of carbon isotope ratios towards significantly more positive $δ^{13}C$ values (-27.5 ± 0.4 ‰ to -20.8 ± 0.5 ‰). After 94 % of γ-HCH removal, no further degradation was observed. In this case, the mass balance of substrate and products and isotope balance remained relatively stable over the course of the experiment indicating that no further degradation of metabolites occurred (Supporting information, Figure S1 D) (Appendix A1).

The carbon isotope enrichment factors were the same for both bacterial strains with $ε_c$ = -1.5 ± 0.1 ‰ and -1.7 ± 0.2 ‰ for *S. indicum* strain B90A and *S. japonicum* strain UT26, respectively (Figure 1, Table 1). Thus, similar reaction mechanisms for γ-HCH biodegradation can be expected for both strains, which are supported by the similarity of γ-HCH degrading *lin* genes in both strains (Lal et al., 2010). The $ε_c$ values calculated for aerobic bacterial strains for γ-HCH were lower as compared to those obtained for anaerobic biodegradation (-3.4 ‰ to -3.9 ‰) (Badea et al., 2009) leading to the assumption that the reaction mechanisms differ for aerobic and anaerobic γ-HCH biodegradation. This is supported by previous studies where it was shown that anaerobic γ-HCH biodegradation is initiated by reductive beta-elimination (Quintero et al., 2005) and aerobic degradation by dehydrochlorination (Lal et al., 2010).

Biodegradation and carbon isotope fractionation of α-HCH

Biodegradation experiments with either *S. indicum* strain B90A or *S. japonicum* strain UT26 and were performed to study carbon isotope fractionation of α-HCH. The degradation of α-HCH was faster in batches inoculated with *S. indicum* strain B90A (1.6 µM h^{-1}) compared to those inoculated with *S. japonicum* strain UT26 (1.1 µM h^{-1}). 1,2,4-TCB and β-pentachlorocyclohexene (β-PCCH) were the main transformation products observed for α-HCH biodegradation with *S. indicum* strain B90A (Figure 2 A). In case of *S. japonicum* strain UT26, 2,5-DCP and 1,2,4-TCB were the main transformation products of α-HCH biodegradation (Supporting information, Figure S2 A) (Appendix A1). The isotope balance of substrate and products was stable during biotransformation of α-HCH with *S. japonicum* strain UT26 (Figure S2 B, supporting information, Appendix A1) indicating that detectable metabolites were not further degraded.

In batches inoculated with *S. indicum* strain B90A, the carbon isotope ratios of α-HCH changed from -27.0 ± 0.1 ‰ to -18.3 ± 0.6 ‰ for 95 % biotransformation (Figure 2 B). In case of *S. japonicum* strain UT26, 96% biotransformation resulted in changes of carbon isotope ratios from -27.7 ± 0.1 ‰ to -21.5 ± 0.5 ‰ (Supporting information, Figure S2 B) (Appendix A1). In the abiotic control batches, the carbon isotope ratios of α-HCH remain stable during the whole experiment (data not shown). In case of *S. indicum* strain B90A however, further degradation occurred as indicated by the decreasing mass balance of substrate and products and the shift of the average carbon isotope composition of α-HCH and its metabolites towards more positive $δ^{13}C$ values (Figure 2A, B).

The degradation experiments with both strains gave similar carbon isotope enrichment factors with $ε_c$ = -1.6 ± 0.3 ‰ and -1.0 ± 0.2 ‰ for *S. indicum* strain B90A and *S. japonicum* strain UT26, respectively (Figure 3A & B, Table 1). The $ε_c$ values of α-HCH when compared with those of γ-HCH exhibited no significant differences (Table 1). This indicates that both HCH isomers were transformed by similar reaction mechanisms. The $ε_c$ values for aerobic were lower than for anaerobic degradation of α-HCH ($ε_c$ = -3.7 ± 0.8 ‰) (Badea et al., 2011) indicating different reaction mechanisms under oxic and anoxic conditions. However, since aerobic biodegradation was significantly faster than anaerobic degradation, other steps than the isotope sensitive carbon bond cleavage might be rate limiting for the overall aerobic biodegradation of α-HCH and therefore mask the carbon isotope fractionation (Aeppli et al., 2009; Nijenhuis et al., 2005).

Temperature can affect biodegradation rates and experiments were conducted to elucidate the dependency of isotope fractionation on degradation rate. Experiments for α-HCH degradation by *S. indicum* strain B90A were conducted at optimal temperature condition at 30°C, and at 20°C and 10°C. The lower temperatures led to lower biodegradation rates but did not change isotope fractionation significantly (Table 1) suggesting the overall degradation kinetic does not affect carbon isotope fractionation in

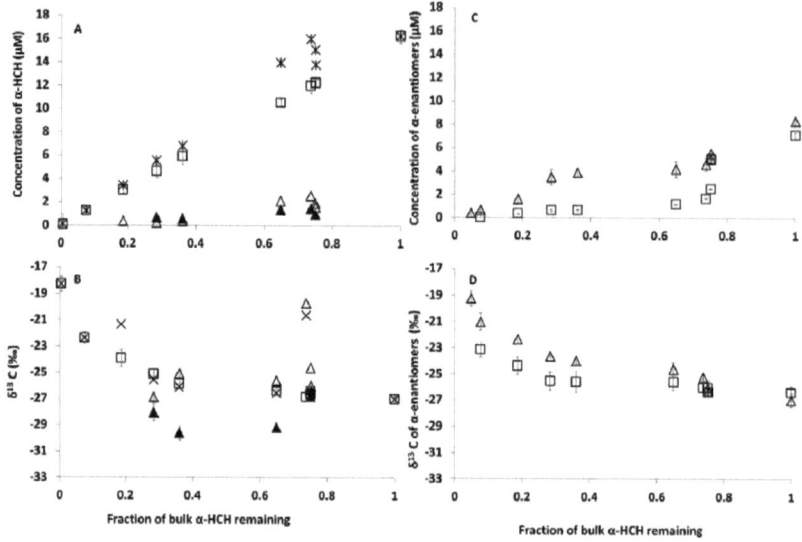

Figure 2: Concentrations (A, C) and carbon isotope ratios (B, D) of α-HCH (A, B) and its enantiomers (C, D) and products during biodegradation by S. indicum strain B90A. (A, B). α-HCH (open squares) and products 1,2,4-TCB (open triangle) and β-pentachlorocyclohexene (β-PCCH) (close triangle). Concentrations (C) and carbon isotope ratios (D) of (-) α-HCH (open square) and (+) α-HCH (open triangle) during biodegradation by S. indicum strain B90A. The sum concentration (A) and average carbon isotope composition (B) of substrate and products (see Material and Methods for calculation) are indicated by X. Error bars indicate the standard deviation of triplicate analysis for isotope analysis

the temperature range between 10 and 30°C. This indicates that the relative low carbon isotope fractionation for aerobic HCH biodegradation seems to mainly caused by the reaction mechanism and the carbon isotope enrichment factor is robust for assessing HCH biodegradation under typical temperature conditions in the environment.

Carbon isotope fractionation of α-HCH enantiomers

The carbon isotope fractionation of (+) and (-) α-HCH enantiomers was studied for aerobic degradation by S. indicum strain B90A and S. japonicum strain UT26. In case of *S. indicum* strain B90A the $\delta^{13}C$ values of the (+) α-enantiomer changed from -27.0 ± 0.8 ‰ to -19.3 ± 0.5 ‰ for 95 % biodegradation, and in the case of (-) α-enantiomer from -26.4 ± 0.1 ‰ to -23.1 ± 0.4 ‰ for 96 % biodegradation (Figure 2 C & D). In the degradation experiment with *S. japonicum* strain UT26 the carbon isotope ratio shifted from -26.8 ± 0.4 ‰ to –24.2 ± 0.3 ‰ for 95 % biotransformation of the (-) α-enantiomer and from -26.1 ± 0.6 ‰ to -19.3 ± 1.0 ‰ for 96 % biotransformation of the (+) α-enantiomer (Supporting information, Figure S2 C & D) (Appendix A1). For anaerobic α-HCH degradation by *C. pasterianum*, partial enantioselectivity was observed (Badea et al., 2011).

For both strains, *S. indicum* strain B90A and *S. japonicum* strain UT26, higher isotope enrichment factors for (+) α-enantiomer (-2.4 ± 0.8 ‰ and -2.5 ± 0.6 ‰, respectively) as compared to the (-) α-enantiomer (-1.0± 0.6 ‰ and -0.7 ± 0.2 ‰, respectively) were obtained (Table 1, Figure 3A & B). The average of the carbon isotope enrichment factors for (+)- and (-)-α-HCH enantiomers were within the statistical uncertainty of the ε_c value of bulk α-HCH. This similarity indicated that the calculation of the enrichment factors for the two enantiomers based on the Rayleigh equation was applicable.

An enantiomer-specific carbon isotope fractionation by both bacterial strains for α-HCH degradation was observed. Lower carbon isotope fractionation of (-) α-enantiomer as compared to (+) α-enantiomer may be caused by different reaction mechanisms for the two enantiomers. However, it was shown that both enantiomers are degraded via the same reaction mechanism (Lal et al., 2010). Thus, different reaction mechanisms can be excluded as reason for the variation in the carbon isotope fractionation of the two enantiomers. The distinction in the carbon isotope fractionation might be caused by different extent of rate limitation preceding the isotope-sensitive step of the biodegradation of the enantiomers. The biodegradation of the (-) α- enantiomer is significantly faster than for the (+) α- enantiomer (Figure 2; Supplementary information, Figure 1). Thus, non-isotope

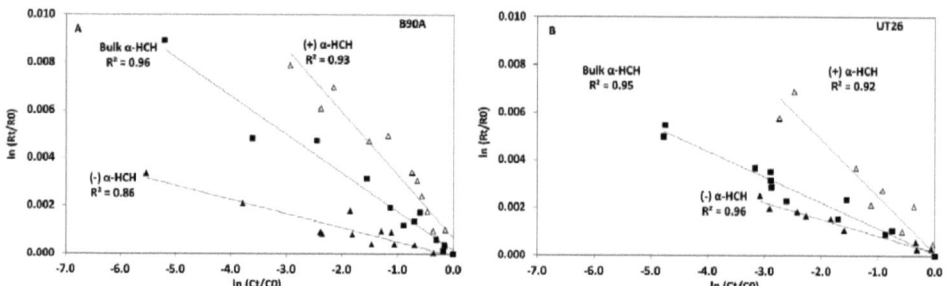

Figure 3: Linearized Rayleigh equation plots showing the carbon isotope fractionation for the biodegradation of bulk α-HCH and its enantiomers by S. indicum strain B90A (A) and S. japonicum strain UT26 (B). Bulk α-HCH (closed square), (-) - (closed triangle) and (+) - (open triangle) α-HCH.

fractionating steps (e.g., substrate uptake into the cell, binding of the substrate to the enzyme) might be more rate-limiting than the isotope-sensitive step of the enzymatic bond cleavage leading to a lower isotope fractionation for the (-) α- enantiomer. The

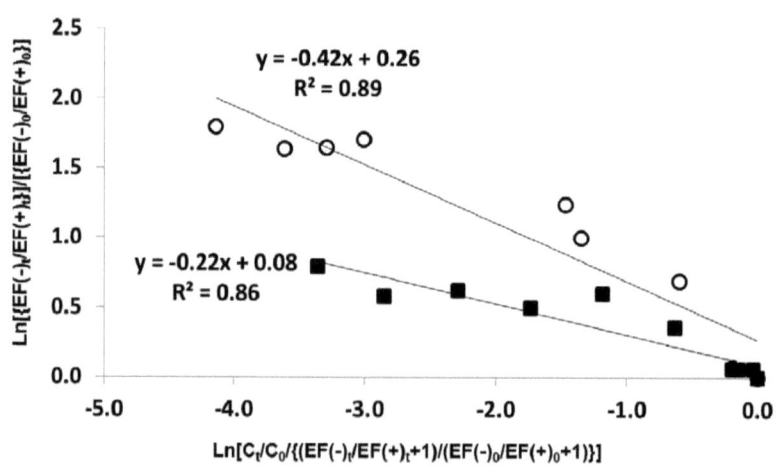

Figure 4: Linearized plots of the general form of the Rayleigh equation (eq. 1) showing the enantiomeric fractionation for the biodegradation of α-HCH by *S. indicum* strain B90A (open circle) and *S. japonicum* strain UT26 (closed square).

slower biodegradation suggested a slower enzymatic turnover of (+) α- enantiomer and this might cause a shift in the rate limitation from non-isotope fractionating, preceding steps towards the isotope-sensitive step of the enzymatic bond cleavage leading to a more pronounced carbon isotope fractionation of (+) α- enantiomer. Indeed, this is reflected in the higher observed isotope fractionation for the (+) compared to the (-) α- enantiomer.

Enantioselective transformation of α-HCH

(-) α-HCH was preferentially degraded with the enantiomeric fraction (EF) (-) values ranging from 0.45 to 0.14 in case of *S. indicum* strain B90A and 0.50 to 0.24 for *S. indicum* strain B90A (Figure 2C, Figure S2C, supporting information, Appendix A1). This may be due to differences in the preferential enzymatic activity of the enantiomers for the two strains. The *lin* genes necessary for aerobic degradation of HCH were initially identified and characterized for *S. japonicum* strain UT26 and were subsequently recovered from *S. indicum* strain B90A as well (Lal et al., 2010) suggesting that similar enzymes are involved in catalysis of α-HCH. However, *S. japonicum* strain UT26 contains LinA while *S. indicum* strain B90A is known to express two copies of *linA* (*linA1* and *linA2*) (Dogra et al., 2004). The amino acid sequences of the products encoded by the *linA1* and *linA2* genes are 88 % identical to each other and 88% (LinA1) and 99 % (LinA2) are similar to the sequence of LinA of *S. japonicum* UT26 (Imai et al., 1991). Preferential degradation of (-) α- enantiomer is known with the LinA1 variant of *S. indicum* B90A(Suar et al., 2005). Similarly, anaerobic experiments performed by Badea *et al,*(2011) showed non-enantioselective transformation of α-HCH by *Clostridium*

Bacteria strain & growth conditions	γ-HCH		Bulk α-HCH		(+) α-HCH	(-) α-HCH
	ε_c (‰)	Rate (µMh^{-1})	ε_c (‰)	Rate (µMh^{-1})	ε_c (‰)	ε_c (‰)
Aerobic						
Sphingobium indicum strain **B90A** 30°C	-1.5 ± 0.1 (n=15)	3.9	-1.6 ± 0.3 (n=10)	1.6	-2.4 ± 0.8 (n=10)	-1.0 ± 0.6 (n=10)
20°C			-1.4 ± 0.6 (n=7)	0.7	-3.3 ± 0.8 (n=7)	-1.0 ± 0.6 (n=7)
10°C			-1.7 ± 0.6 (n=7)	0.6	-2.7 ± 1.0 (n=7)	-0.7 ± 0.2 (n=7)
Sphingobium japonicum strain **UT26** 30°C	-1.7 ± 0.2 (n=19)	1.2	-1.0 ± 0.2 (n=13)	1.1	-2.5 ± 0.6 (n=11)	-0.7 ± 0.2 (n=10)
Anaerobic						
Clostridium pasterianum			-3.7 ± 0.8*			
Desulfovibrio gigas	-3.9 ± 0.6$					
Desulfococcus mulitvorans	-3.4 ± 0.5$					

Table 1: Carbon isotope enrichment factors (ε_c) of γ-HCH, α-HCH and its enantiomers and related degradation rates. In brackets: number of samples. *Badea et al. (2011); $Badea et al. (2009).

pasterianum indicating the potential of the distinction for aerobic and anaerobic α-HCH biodegradation using EF (+) or EF (-) values.

The most general form of the Rayleigh equation (eq. 2) was applied for calculating enantiomeric enrichment factors (ε_e) for aerobic α-HCH biodegradation (Figure 4). This equation is applicable to describe fractionation processes of enantiomers exhibiting similar reaction kinetics(Gasser et al., 2012) . For aerobic α-HCH

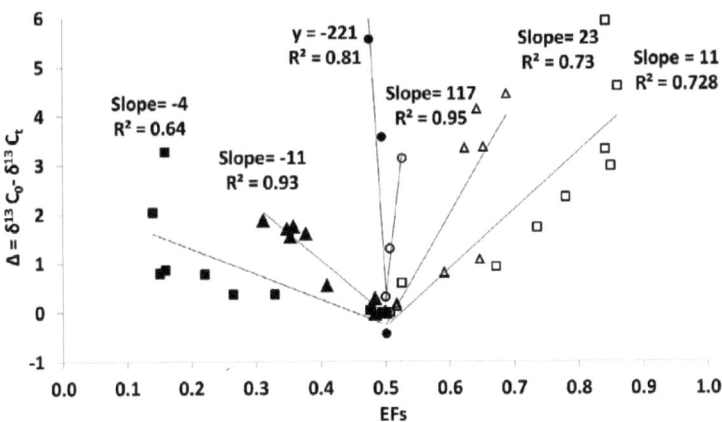

Figure 5: Comparison of carbon isotope discrimination ($\Delta = \delta_t - \delta_0$) vs. enantiomeric fractions (EFs) for (-) α-HCH (EF (-)) (closed symbols) and (+) α-HCH (EF (+)) (open symbols) for anaerobic biodegradation (circles) and aerobic biodegradation by B90A (squares) and UT26 (triangles).

biodegradation, the enantiomeric fractionation could be described by the Rayleigh equation, and the ε_e values calculated for S. indicum strain B90A and S. japonicum strain UT26 were -43 ± 15 % and -22 ± 6 %, respectively. The correlation coefficients (R^2) of 0.89 and 0.86 for S. indicum strain B90A and S. japonicum strain UT26, respectively, suggested slightly different reaction kinetics of the two enantiomers reducing the fit of the linear correlation of the Rayleigh equation plots (Figure 4).

The combined isotope and enantiomeric fractionation has the potential to differentiate aerobic and anaerobic biodegradation of α-HCH. In order to combine both approaches, EF(+) was plotted against the carbon isotope discrimination ($\Delta = \delta_t - \delta_0$) of the (+) α- enantiomer as well as EF(-) against the carbon isotope discrimination of (-) α-enantiomer (Figure 5) and compared to previous of anaerobic α-HCH biodegradation (Badea et al., 2011). Slopes for the linear regressions of EFs vs. carbon isotope discrimination were for aerobic biodegradation -4 and -11 for the (-) - enantiomer and +11 to +23, for the (+) - enantiomer for S. indicum strain B90A and S. japonicum strain UT26, respectively. For anaerobic biodegradation, slopes were obtained with -221 and +117 for the (-) - and (+) - enantiomer, respectively. The different trends indicate the potential for distinguishing aerobic and anaerobic biodegradation of α-HCH in the environment based on isotope and enantiomeric fractionation however needed to be confirmed and validated by investigating further aerobic and anaerobic microbial cultures.

Implications for Environmental Studies

In this study, we present the combination of isotopic and enantiomeric fractionation providing a novel approach for monitoring of *in situ* biodegradation incorporating the isotope signatures of isomers and enantiomers of HCH as well as enantiomer selective degradation. We determined the isotope enrichment factors for the aerobic biodegradation of α- and γ-HCH for the first time allowing a quantification of HCH biodegradation in oxic compartments. Additionally, the differences in isotope fractionation between aerobic and anaerobic degradation indicates that CSIA can be used to evaluate degradation pathways.

Enantiomer fractionation provides another indicator for *in situ* biodegradation of α-HCH similar to previous observations (Harner et al., 2000b; Helm et al., 2000; Law et al., 2004). By applying the Rayleigh equation, we obtained the enantiomer enrichment factors for aerobic α-HCH biodegradation providing a quantitative framework, complementary to the approach via CSIA, which can be determined by common GC analysis. Moreover, this approach might also be applicable for other enantiomeric pollutants as it was depicted for *O*-desmethylvenlafaxine (Gasser et al., 2012).

Based on the recently developed ESIA approach (Badea et al., 2011), we observed significant carbon isotope fractionation, which was different for the each of the α-HCH enantiomers during aerobic biodegradation. Since each individual enantiomer may have different rates of degradation, quantification of *in situ* biodegradation based on enantiomer-specific isotope analysis allows a more refined assessment of contaminated field site with chiral pollutants as it was confirmed by a recent field study(Milosevic et al., 2012). This enantiomer-specific analysis is especially important in case one enantiomer is subject to biodegradation while the other persists.

Finally, the combination of enantiomer and carbon isotope fractionation, in a two-dimensional approach, using values of EF(+) and EF(-) in correlation with the stable carbon isotope discrimination ($\Delta\delta^{13}C$) of the α-HCH enantiomers should allow to trace degradation pathways of α-HCH in the environment (Figure 4). The specific ranges for the pathways obtained in our study can, thereby, be used as reference for evaluating field data in order to determine the relevance of aerobic and anaerobic α-HCH degradation at contaminated sites.

In summary, our study provides a concept allowing the use of enantiomer fractionation, CSIA and ESIA for a complementary and thus comprehensive assessment of *in situ* degradation of α-HCH, but also other chiral contaminants. Thus, we provide proof of principle for the theoretical framework of a multiple lines of evidence approach for i) providing evidence of degradation, ii) for distinguishing pathways and iii) for quantifying HCH degradation at contaminated field sites. However, as other degradation pathways such as dehydrogenation and oxidation may be responsible for HCH biodegradation in a field sites in addition to the so far known reductive beta-elimination or dehydrogenation, reference culture experiments with pure cultures having these respective degradation pathways should be performed. Additionally, the applicability of the combined enantiomer fractionation, CSIA and ESIA approach needs to be tested in field studies.

Associated content

Supporting information
This material is attached in Chapter A1 Appendix

Acknowledgements
We thank to Mathias Gehre, Ursula Gunther and Falk Bratfisch for the technical support of the isotope analysis. We also thank to Prof. Rup Lal and Prof. Jiri Damborský for providing bacterial cultures. Safdar Bashir was supported by University of Agriculture Faisalabad, Pakistan and the Helmholtz Impulse and Networking Fund through Helmholtz Interdisciplinary Graduate School for Environmental Research (HIGRADE).

References

Aeppli C., Berg M., Cirpka O.A., Holliger C., Schwarzenbach R.P., Hofstetter T.B. (2009) Influence of mass-transfer limitations on carbon isotope fractionation during microbial dechlorination of trichloroethene. Environmental Science & Technology, 43(23):8813-8820.

Badea S.L., Vogt C., Weber S., Danet A.F., Richnow H.H. (2009) Stable isotope fractionation of gamma-Hexachlorocyclohexane (Lindane) during reductive dechlorination by two strains of sulfate-reducing bacteria. Environmental Science & Technology, 43(9):3155-316.

Badea S.L., Vogt C., Gehre M., Fischer A., Danet A.F., Richnow H.H. (2011) Development of an enantiomer-specific stable carbon isotope analysis (ESIA) method for assessing the fate of α-hexachlorocyclo-hexane in the environment. Rapid Communications in Mass Spectrometry, 25(10):1363-1372.

Bhatt P., Kumar M.S., Chakrabarti T. (2009) Fate and degradation of POP-hexachlorocyclohexane. Critical Reviews in Environmental Science & Technology, 39(8):655-695.

Breivik K., Pacyna J.M., Münch J. (1999) Use of α-, β-and γ-hexachlorocyclohexane in Europe, 1970-1996. Science of The Total Environment, 239(1-3):151-163.

Butte W., Fox K., Zauke G.P. (1991) Kinetics of bioaccumulation and clearance of isomeric hexachlorocyclohexanes. Science of The Total Environment, 109:377-382.

Canton J., Greve P., Slooff W., Van Esch G. (1975) Toxicity, accumulation and elimination studies of α-hexachlorocyclohexane (α-HCH) with freshwater organisms of different trophic levels. Water Research, 9(12):1163-1169.

Cichocka D., Imfeld G., Richnow H.-H., Nijenhuis I. (2008) Variability in microbial carbon isotope fractionation of tetra-and trichloroethene upon reductive dechlorination. Chemosphere, 71(4):639-648.

Coplen T.B. (2011) Guidelines and recommended terms for expression of stable-isotope-ratio and gas-ratio measurement results. Rapid Communications in Mass Spectrometry, 25(17):2538-2560.

Dogra C., Raina V., Pal R., Suar M., Lal S., Gartemann K.H., Holliger C., van der Meer J.R., Lal R. (2004) Organization of lin genes and IS6100 among different strains of hexachlorocyclohexane-degrading *Sphingomonas paucimobilis*:

evidence for horizontal gene transfer. Journal of Bacteriology, 186(8):2225-2235.

Elsner M., McKelvie J., Couloume G.L., Lollar B.S. (2007) Insight into methyl tert-butyl ether (MTBE) stable isotope fractionation from abiotic reference experiments. Environmental Science & Technology, 41(16):5693-5700.

Endo R., Ohtsubo Y., Tsuda M., Nagata Y. (2006) Growth inhibition by metabolites of γ-hexachlorocyclohexane in *Sphingobium japonicum* UT26. Bioscience, Biotechnology, and Biochemistry, 70(4):1029-1032.

Garrison A.W. (2006) Probing the enantioselectivity of chiral pesticides. Environmental Science & Technology, 40(1):16-23.

Gasser G., Pankratov I., Elhanany S., Werner P., Gun J., Gelman F., Lev O. (2012) Field and laboratory studies of the fate and enantiomeric enrichment of venlafaxine and O-desmethylvenlafaxine under aerobic and anaerobic conditions. Chemosphere, 88(1):98-105.

Geueke B., Garg N., Ghosh S., Fleischmann T., Holliger C., Lal R., Kohler H.-P.E. (2012) Metabolomics of hexachlorocyclohexane (HCH) transformation: ratio of LinA to LinB determines metabolic fate of HCH isomers. Environmental Microbiology, 15(4):1040-1049.

Harner T., Wiberg K., Norstrom R. (2000a) Enantiomer fractions are preferred to enantiomer ratios for describing chiral signatures in environmental analysis. Environmental Science & Technology, 34(1):218-220.

Harner T., Jantunen L.M., Bidleman T.F., Barrie L.A., Kylin H., Strachan W.M., Macdonald R.W. (2000b) Microbial degradation is a key elimination pathway of hexachlorocyclohexanes from the Arctic Ocean. Geophysical Research Letters, 27(8):1155-1158.

Hegeman W.J., Laane R.W. (2001) Enantiomeric enrichment of chiral pesticides in the environment. Reviews of Environmental Contamination and Toxicology, 173 173:85-116.

Helm P.A., Diamond M.L., Semkin R., Bidleman T.F. (2000) Degradation as a loss mechanism in the fate of α-hexachlorocyclohexane in Arctic watersheds. Environmental Science & Technology, 34(5):812-818.

Hofstetter T.B., Berg M. (2011) Assessing transformation processes of organic contaminants by compound-specific stable isotope analysis. TrAC Trends in Analytical Chemistry, 30(4):618-627.

Imai R., Nagata Y., Fukuda M., Takagi M., Yano K. (1991) Molecular cloning of a *Pseudomonas paucimobilis* gene encoding a 17-kilodalton polypeptide that eliminates HCl molecules from γ-hexachlorocyclohexane. Journal of Bacteriology, 173(21):6811-6819.

Imai R., Nagata Y., Senoo K., Wada H., Fukuda M., Takagi M., Yano K. (1989) Dehydrochlorination of γ-hexachlorocyclohexane(γ-BHC) by γ-BHC-assimilating *Pseudomonas paucimobilis*. Agricultural and Biological Chemistry, 53(7):2015-2017.

Kallenborn R. (2001) Chiral environmental pollutants: trace analysis and ecotoxicology. Springer.

Kohler H.P.E., Angst W., Giger W., Kanz C., Muller S., Suter M.J.F. (1997) Environmental fate of chiral pollutants ndash the necessity of considering stereochemistry. CHIMIA International Journal for Chemistry, 51(12):947-951.

Kumari R., Subudhi S., Suar M., Dhingra G., Raina V., Dogra C., Lal S., van der Meer J.R., Holliger C., Lal R. (2002) Cloning and characterization of lin genes responsible for the degradation of hexachlorocyclohexane isomers by *Sphingomonas paucimobilis* strain B90. Applied and Environmental Microbiology, 68(12):6021-6028.

Lal R., Pandey G., Sharma P., Kumari K., Malhotra S., Pandey R., Raina V., Kohler H.P.E., Holliger C., Jackson C., Oakeshott J.G. (2010) Biochemistry of microbial degradation of Hexachlorocyclohexane and prospects for bioremediation. Microbiology and Molecular Biology Reviews, 74(1):58-80.

Law S.A., Bidleman T.F., Martin M.J., Ruby M.V. (2004) Evidence of enantioselective degradation of α-hexachlorocyclohexane in groundwater. Environmental Science & Technology, 38(6):1633-1638.

Li Y. (1999) Global technical hexachlorocyclohexane usage and its contamination consequences in the environment: from 1948 to 1997. Science of The Total Environment, 232(3):121-158.

Li Y.F., Macdonald R.W. (2005) Sources and pathways of selected organochlorine pesticides to the Arctic and the effect of pathway divergence on HCH trends in biota: a review. Science of The Total Environment, 342(1-3):87-106.

Mariotti A., Germon J., Hubert P., Kaiser P., Letolle R., Tardieux A., Tardieux P. (1981) Experimental determination of nitrogen kinetic isotope fractionation:

some principles; illustration for the denitrification and nitrification processes. Plant and Soil, 62(3):413-430.

Milosevic N., Qiu S., Elsner M., Einsiedl F., Maier M., Bensch H., Albrechtsen H.-J., Bjerg P.L. (2012) Combined isotope and enantiomer analysis to assess the fate of phenoxy acids in a heterogeneous geologic setting at an old landfill. Water Research, 47(2):637−649.

Nagasawa S., Kikuchi R., Matsuo M. (1993a) Indirect identification of an unstable intermediate in γ-HCH degradation by *Pseudomonas paucimobilis* UT26. Chemosphere, 26(12):2279-2288.

Nagasawa S., Kikuchi R., Nagata Y., Takagi M., Matsuo M. (1993b) Aerobic mineralization of γ-HCH by *Pseudomonas paucimobilis* Ut26. Chemosphere, 26(9):1719-1728.

Nagata Y., Futamura A., Miyauchi K., Takagi M. (1999) Two different types of dehalogenases, LinA and LinB, involved in γ-Hexachlorocyclohexane degradation in *Sphingomonas paucimobilis* UT26 are localized in the periplasmic space without molecular processing. Journal of Bacteriology, 181(17):5409-5413.

Nijenhuis I., Andert J., Beck K., Kästner M., Diekert G., Richnow H.-H. (2005) Stable isotope fractionation of tetrachloroethene during reductive dechlorination by *Sulfurospirillum multivorans* and *Desulfitobacterium* sp. strain PCE-S and abiotic reactions with cyanocobalamin. Applied and Environmental Microbiology, 71(7):3413-3419.

Quintero J.C., Moreira M.T., Feijoo G., Lema J.M. (2005) Anaerobic degradation of hexachlorocyclohexane isomers in liquid and soil slurry systems. Chemosphere, 61(4):528-536.

Quintero J.C., Moreira M.T., Lema J.M., Feijoo G. (2006) An anaerobic bioreactor allows the efficient degradation of HCH isomers in soil slurry. Chemosphere, 63(6):1005-1013.

Raina V., Suar M., Singh A., Prakash O., Dadhwal M., Gupta S.K., Dogra C., Lawlor K., Lal S., van der Meer J.R. (2008) Enhanced biodegradation of hexachlorocyclohexane (HCH) in contaminated soils via inoculation with *Sphingobium indicum* B90A. Biodegradation, 19(1):27-40.

Rosell M., Finsterbusch S., Jechalke S., Hübschmann T., Vogt C., Richnow H.H. (2009) Evaluation of the effects of low oxygen concentration on stable isotope fractionation during aerobic MTBE biodegradation. Environmental Science & Technology, 44(1):309-315.

Solomon L.M., Fahrner L., West D.P. (1977) Gamma benzene hexachloride toxicity: a review. Archives of Dermatology, 113(3):353-357.

Suar M., Hauser A., Poiger T., Buser H.R., Müller M.D., Dogra C., Raina V., Holliger C., van der Meer J.R., Lal R. (2005) Enantioselective transformation of α-hexachlorocyclohexane by the dehydrochlorinases LinA1 and LinA2 from the soil bacterium *Sphingomonas paucimobilis* B90A. Applied and Environmental Microbiology, 71(12):8514-8518.

Swartjes F. A. (2011). Dealing with contaminated sites (pp. 1079-1104). Netherlands: Springer.

Thullner M., Centler F., Richnow H.H., Fischer A. (2012) Quantification of organic pollutant degradation in contaminated aquifers using compound-specific stable isotope analysis - review of recent developments. Organic Geochemistry, 42(12):1440-1460.

Trantírek L., Hynková K., Nagata Y., Murzin A., Ansorgová A., Sklenář V., Damborský J. (2001) Reaction mechanism and stereochemistry of γ-hexachlorocyclohexane dehydrochlorinase LinA. Journal of Biological Chemistry, 276(11):7734-7740.

Vijgen J., Abhilash P., Li Y.F., Lal R., Forter M., Torres J., Singh N., Yunus M., Tian C., Schäffer A. (2011) Hexachlorocyclohexane (HCH) as new Stockholm convention POPs-a global perspective on the management of Lindane and its waste isomers. Environmental Science and Pollution Research, 18(2):152-162.

Voldner E.C., Li Y.F. (1995) Global usage of selected persistent organochlorines. Science of The Total Environment, 160:201-210.

Wiberg K., Harner T., Wideman J.L., Bidleman T.F. (2001) Chiral analysis of organochlorine pesticides in Alabama soils. Chemosphere, 45(6):843-848.

Willett K.L., Ulrich E.M., Hites R.A. (1998) Differential toxicity and environmental fates of hexachlorocyclohexane isomers. Environmental Science & Technology, 32(15):2197-2207.

Chapter : 3.2 Results

3.2

Anaerobic Biotransformation of Hexachlorocyclohexane Isomers by *Dehalococcoides* spp.

Safdar Bashir, Kevin Kuntze, Carsten Vogt and Ivonne Nijenhuis

Department of Isotope Biogeochemistry, Helmholtz Centre for Environmental Research – UFZ, Permoserstraße 15, 04318, Leipzig,
Germany

Abstract

The biotransformation of hexachlorocyclohexane (HCH) by two *Dehalococcoides mccartyi* strains and an enrichment culture was investigated. Both *D. mccartyi* strains preferentially degraded γ-HCH over α-HCH and δ-HCH isomers while β-HCH biotransformation was not significant. *D. mccartyi* strain 195 was capable of growth with γ-HCH as terminal electron acceptor, however, only after initial cultivation with tetrachloroethene (PCE). The enrichment culture preferentially transformed γ-HCH over the δ-HCH, β-HCH and α-HCH isomers. Major observed metabolite was tetrachlorocyclohexene with monochlorobenzene and benzene as end products. Carbon stable isotope analysis confirmed the similarity in degradation pathways under anoxic conditions with the enrichment factor ε_c = -5.5 ± 0.8 ‰ for *D. mccartyi* strain 195, ε_c = -3.1 ± 0.4 ‰ for the enrichment culture and ε_c = -4.1 ± 0.6 ‰ for *Clostridium pasteurianum* DSMZ 525, as a reference.

3.2.1 Introduction

γ-hexachlorocyclohexane (γ-HCH, Lindane) was one of the most popular agricultural insecticides deployed from early 1950 until early 1990 which global production resulted in more than 4 million tons of HCH waste (Vijgen et al., 2011). Technical-grade HCH consists of five main isomers: α (55-80 %), β (5-14 %), γ (8-15 %), δ (2-16 %) and ε (1-5 %) differing in spatial orientation of chlorine atoms around the cyclohexane ring resulting in different stabilities and physico-chemical properties (Willett et al., 1998). The environmentally most significant isomers present in soil and groundwater are α-, β-, γ- and δ-HCH (Buser and Mueller, 1995; Popp et al., 2000; Ricking and Schwarzbauer, 2008). All HCH isomers are toxic, considered to be carcinogenic and have bioaccumulation potential. Consequently, HCHs pose a substantial risk for human health and ecosystems and there is an urgent need for sustainable approaches to remove them from the environment (Phillips et al., 2005; Walker et al., 1999) e.g. via bioremediation.

Studies on microbial degradation of HCH included pure and enrichment cultures under both aerobic and anaerobic conditions (Bachmann et al., 1988; Lal et al., 2010; Mehboob et al., 2013; Nagasawa et al., 1993; Sahu et al., 1995). Most bacteria able to degrade HCH under aerobic conditions belong to the family *Sphingomonadaceae*, and the key enzymes and encoding genes have already been identified and characterized (for a review see (Lal et al., 2010). A dehydrochlorination reaction initiates the aerobic degradation resulting in pentachlorocyclohexene (PCCH) as the first metabolite and two further dehydrochlorination reactions produce two putative intermediates, 1,3,4,6-tetrachloro-1,4-cyclohexadiene (1,3,4,6-TCDN) and 2,5-dichloro-2,5-cyclohexadiene-1,4-diol (2,5-DDOL), resulting either in the end products 1,2,4-trichlorobenzene (1,2,4-TCB) and 2,5-dichloropenols (2,5-DCP) by abiotic processes, or a complete mineralization (Lal et al., 2010).

Under anaerobic conditions, as found in e.g. contaminated aquifers or sediments, dichloroelimination and dehydrochlorination seem to be the main processes during transformation of HCH isomers via tetrachlorocyclohexene (TCCH) into lower chlorinated products, e.g. dichlorobenzene (DCB), monochlorobenzene, and benzene (Lal et al., 2010; Middeldorp et al., 1996). For example, cell suspensions of different sulfate reducing bacteria were shown to dehalogenate γ-HCH (Boyle et al., 1999), and granular sludge and soils were observed to dehalogenate α-, β-, γ- and δ-HCH (Middeldorp et al., 1996; Van Eekert et al., 1998) MCB and benzene were end products of HCH transformations in these cultures. Cell suspensions of *Clostridium* spp. were observed to transform the α- and γ-HCH isomers (Heritage and MacRae, 1977; Jagnow et al., 1977b; MacRae et al., 1969). In these studies, the dechlorination of HCH seemingly proceeds through co-metabolic reactions which are unspecific and often slow.

A reductive dechlorination of β-HCH to MCB and benzene was shown with an anaerobic *Dehalobacter* co-culture, solely growing with H_2 as electron donor and β - HCH as electron acceptor, indicating that dechlorination can be a respiratory process (Doesburg et al., 2005). Additionally, Elango et al., 2010 described an enrichment culture which was growing with hydrogen and γ-HCH as sole electron donor and acceptor, respectively, producing also mainly benzene and MCB as end products. Thus far, though, no isolate has been shown capable of HCH transformation coupled to growth.

Over the last decades, strains of the genus *Dehalococcoides* have received significant attention because of their ability to reductively dehalogenate common groundwater contaminants such as chlorinated ethenes, ethanes and benzenes due to their extensive genomic inventory of putative reductive dehalogenases, mostly with unknown function (Kaufhold et al., 2013; Krajmalnik-Brown et al., 2007; Löffler et al., 2013; Pöritz et al., 2013; Seshadri et al., 2005). Recently, (Kaufhold et al., 2013) reported the transformation of γ-HCH isomer to MCB by a highly enriched *D. mccartyi* strain BTF08-containing culture as well as by *D. mccartyi* strain 195. However, biotransformation of other HCH isomers and utilization as growth substrate was not tested.

Therefore, this study focused on the biotransformation of HCHs by *D. mccartyi* strain 195 and strain BTF08 and related growth. The degradation pathway was investigated by analysing metabolites and end products. Furthermore, the stable carbon isotope composition of HCH isomers was monitored during biotransformation were assessed and compared to the co-metabolic HCH degrading *Clostridium pasteurianum* (Jagnow et al., 1977a) as well as to a γ-HCH degrading enrichment culture to investigate the similarity in initial reaction.

3.2.2 Materials and methods
Chemicals

HCH isomers [γ-HCH (analytical purity, 97 %), α-HCH (99 %), β-HCH (99 %), δ-HCH (99.5 %)], benzene and MCB (≥ 99 %), hexachlorobenzene (99 %) and toluene (≥9 9 %) were purchased from Sigma Aldrich (Germany). Pentane (analytical purity > 99 %) was obtained from Carl Roth (Germany). Trichlorobenzenes were purchased from Fluka (Germany) and Tetrachloroethene (PCE) was purchased from Merck (Germany).

Bacterial cultures and growth conditions

Dehalococcoides mccartyi strains 195 (Maymo-Gatell, 1997) and BTF08 were obtained from the laboratory culture collection. *Clostridium pasteurianum* DSMZ 525 was purchased from the Leibniz Institute DSMZ - German Collection of Microorganisms and Cell Cultures, Braunschweig, Germany. *D. mccartyi* strain 195 was grown in an anaerobic medium as previously described and was incubated at 34°C and 150 rpm shaking (Cichocka et al., 2008). Growth medium and incubation conditions for *Dehalococcoides mccartyi* strain BTF08 were identical as described previously (Kaufhold et al., 2013) with the exception of shaking continuously at 120 rpm. For both strains, H_2 (0.5 bar) was used as electron donor and acetate as carbon source (5 mM). Growth conditions for *Clostridium pasteurianum* DSM 525 were used as described by (Badea et al., 2011).

Degradation of HCH isomers

Batch culture experiments were performed in 50 ml glass serum bottles filled with 25 ML of the respective medium. Parallel batches (in total 12) were prepared for each strain of which three were kept as controls without inoculum. For experiments performed with PCE pre-grown cultures the setup was the same but cultures were initially amended twice with PCE (100 µM). After complete transformation of PCE was observed, HCH isomers (prepared in acetone stock solution) were spiked at a final concentration of 25 µM in each batch. Sampling was done periodically to assess the production of expected metabolites and each batch was sacrificed at different extents of biodegradation. HCH extraction of the complete bottle with n-pentane (1 ml) for quantification, isotope and metabolite analysis was done as described previously (Bashir et al., 2013). To assess growth, cell numbers were determined microscopically as described previously (Adrian et al., 2007).

Microcosms and enrichment cultures

Groundwater was taken from an HCH-contaminated aquifer located in Bitterfeld-Wolfen (Germany) previously investigated for natural attenuation of chlorobenzenes (Schmidt et al., 2014). Microcosms were prepared as described previously (Nijenhuis et al., 2007). Briefly, quartz sand (1 g) was added in a 120 ML serum bottle and γ-HCH was spiked at a final concentration of 100 µM as electron acceptor and the acetone was subsequently evaporated. Groundwater (50 ML) was added in each bottle under anoxic conditions by using an anoxic chamber (Coy laboratories, USA). The bottles were closed with Teflon® coated septum and crimped. Three sets of treatments were prepared: H_2 (0.5 bar)/Na-acetate (5 mM), lactate (5 mM) or formate (5 mM), respectively, as energy and carbon source, and γ-HCH as potential electron acceptor. All bottles were additionally amended with $NaHCO_3$ (11.9 mM), Na_2S (0.32 mM), , and vitamins (Maymo-Gatell, 1997). In total, five bottles were prepared from which two were kept as killed controls (autoclaved at three consecutive days). A separate set of three bottles without added electron donor was kept as live control. All microcosms were incubated at 20°C without shaking. In the enrichment steps (in total three transfers), active microcosms were transferred (5 % (v/v) inoculum) into medium described previously (Zinder, 1998) and γ-HCH (0.1 M stock solution prepared in acetone) was used as potential electron acceptor at a final concentration of 25 µM.

Molecular biology methods

To analyze the microbial community 2 ML of the respective culture was filtered (0.2 µm) and the collected genomic DNA was extracted with the DNeasy Tissue kit (QIAGEN, Hilden, Germany) following the manufacturer's instructions. The DNA was eluted in 40 µL DNase-free water. The presence of 16S rDNA genes of *Dehalococcoides* was tested by using the primer combination DHC1/DHC1377 and DHC774/DHC1212, respectively, as described by (Hendrickson et al., 2002). PCR amplifications were performed using the GoTaq® Green Master Mix (Promega, USA). The PCR mixtures contained PCR buffer, 1.5 mM $MgCl_2$, deoxynucleoside triphosphate (200 µM each), the respective primers (20 pmol), Taq polymerase (2.5 U), and the extracted DNA as template. The following PCR thermocycling program (Mastercycler, Eppendorf, Germany) was used: 10 min of denaturation at 95°C, followed by either 30 or 35 cycles of 1 min at 95°C, 1 min at 55°C, and 1:30 min at 72°C and finally cooling at 4°C as described previously (Hendrickson et al., 2002).

Analytical approaches

Transformation of HCH isomers was monitored by analyzing tri-, di-, monochlorobenzene and benzene as possible metabolites/end-products. Liquid samples (1 ML) were transferred into 10 ML vials containing 0.5 ML of a saturated Na_2SO_4/H_2SO_4 solution (pH 1) to inhibit microbial activity. Samples were analyzed using gas chromatography (Agilent 6890; Agilent Technologies, Palo Alto, CA) with flame ionization detection and a Rtx-VMS column (Restek, Bad Homburg, Germany) with a length of 30 m, an inner diameter of 0.25 mm and a film thickness of 1.4 µm. The temperature program was used as described previously (Kaufhold et al., 2013). For quantifying remaining substrate concentration and to identify the metabolites, a gas chromatograph (GC) (7890A, Agilent Technologies, Palo, USA) coupled to a mass spectrometer (MS) (5975C, Agilent Technologies, Palo, USA) was used. HCH isomers and their metabolites were separated by BPX-5 capillary column (30m x 0.25mm x 0.25µM; SGE, Darmstadt, Germany). The temperature program was used as described previously (Badea et al., 2009).

A gas chromatograph-combustion-isotope ratio mass spectrometer (GC-C-IRMS) was used to analyze the stable carbon isotope ratios of HCH isomers. The system contained a GC (6890, Agilent Technologies, Palo, USA) coupled with Conflow III interface (Thermo Fisher Scientific, Bremen, Germany) to an IRMS (MAT252, Thermo Fisher Scientific, Bremen, Germany) as described previously (Badea et al.,

2009; Badea et al., 2011). Three µL aliquots of *n*-pentane extract were injected with a split ration of 1:3 for samples with HCH concentration above and splitless for samples below 2 µM. All samples were measured in at least triplicates.

The carbon isotope enrichment factor (ε_c) was determined by the logarithmic form of the Rayleigh equation (Mariotti et al., 1981), plotting ln(C_t/C_0) versus ln[($\delta_t^{13}C+1$)/($\delta_0^{13}C+1$)] and obtaining ε_c from the slope of the linear regression (m = ε_c).

$$ln\left(\frac{\delta_t^{13}C+1}{\delta_0^{13}C+1}\right) = \varepsilon_c ln\left(\frac{C_t}{C_0}\right)$$

The error of the isotope enrichment factors is reported as 95 % confidence interval (CI) determined by a regression curve analysis.

3.2.3 Results and discussion

Biotransformation of HCH isomers by *Dehalococcoides mccartyi* strains

Biotransformation of single α-, β-, γ- and δ-HCH isomers by *D. mccartyi* strain 195 and strain BTF08 was tested using a pre-culture grown on tetrachloroethylene (PCE) and vinyl chloride (VC), respectively. After 60 and 120 days of incubation low amounts of the expected metabolite MCB (up to 10 µM for γ-HCH) were detected for strain 195 and strain BTF08, respectively (Figure S1, supporting information, Appendix A2), as well as traces of TeCCH and benzene (data not shown). Degradation rates of different isomers were γ-HCH > α-HCH > β-HCH/δ-HCH and therefore in similar order reported previously for co-metabolic reactions (Jagnow et al., 1977a). No production of metabolites was observed in abiotic controls (Figure S1, supporting information, Appendix A2).

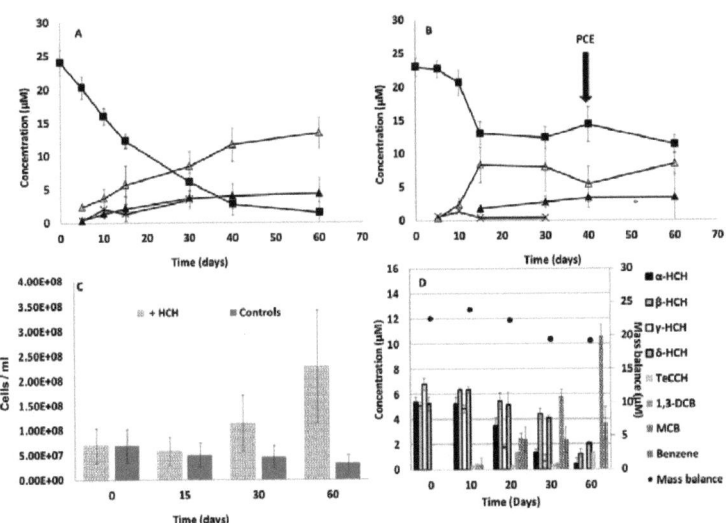

Figure 1: Biodegradation of γ-HCH (A) and δ-HCH (B), γ-HCH- related growth (C) and biodegradation of HCH mixtures by D. mccartyi strain 195 pre-grown with PCE. Concentrations of γ-HCH & δ-HCH (filled squares in panel A and B respectively) and the observed products MCB (open triangles), benzene (filled triangles) and TeCCH (X). C) Cell numbers of D. mccartyi strain 195 with (lighter grey) and without (darker grey) addition of PCE, H2 and acetate D) Changes in concentration of the α-, β-, γ-, and δ-HCH isomers and production of intermediates by a γ-HCH grown D. mccartyi strain 195 culture.

In the experiment with single isomers only a low amount of HCH was transformed over a relatively long incubation time, therefore, to test if the biotransformation of γ-HCH was metabolic or co-metabolic, batch experiments were performed in which D. mccartyi strain 195 was grown initially with PCE (200 µmol L^{-1}) as terminal electron acceptor followed by the addition of the respective single α-, β-, γ- or δ-HCH isomer (25 µmol L^{-1}). No metabolites were observed for α- or β-HCH whereas a complete reduction of γ-HCH was observed with the appearance of TeCCH initially and then MCB and benzene were observed simultaneously within 60 days (Figure 1A). The biodegradation activity in δ-HCH spiked cultures stopped after 56 % of substrate degraded and was not improved by further addition of carbon source (acetate), electron donor (H$_2$) and PCE (Figure 1B). The growth of D. mccartyi strain 195 was further monitored in γ-HCH spiked batches; cell numbers increased in cultures compared to controls without addition of γ-HCH (Figure 1C). Kaufhold et al. (2013) presented already the γ-HCH biotransformation potential of strain 195, however, no significant 16S rRNA gene copy number increase was observed with only 25 % γ-HCH was transformation, but in our study complete transformation of γ-HCH was observed. Our study demonstrates for the first time the capability of strain 195 to use γ-HCH as a substrate for growth. However, the biotransformation seems to be dependent on an initial growth with PCE as terminal electron acceptor potentially due to energy needed for *de novo* synbook of the appropriate dehalogenases or other growth factors. The usage of H_2 as electron donor indicates that dechlorination of by *D. mccartyi* strain 195 is a respiratory process, as recently shown for a γ-HCH dechlorinating anaerobic enrichment culture (Elango et al., 2010) and for a β-HCH dechlorinating *Dehalobacter* co-culture (Doesburg et al., 2005). The *Dehalobacter* co-culture dechlorinates preferentially α- and β-HCH and thus differs from strain 195, with its preference for γ-HCH indicating that specific enzymes are involved in the transformation of the respective isomers.

Since the HCHs are usually present in the environment as mixtures of isomers and to test if the biotransformation of α-, β- and δ-HCH isomers was stimulated by γ-HCH addition, PCE pre-grown batches (triplicates) were initially spiked with γ-HCH (5 µM). After complete transformation of γ-HCH, a mixture of HCH isomers (α, β, γ and δ-HCH, 25 µM in total) was added. A decrease in concentration of all HCH isomers was observed with a balanced increase of the metabolites TeCCH, MCB and benzene as well as traces of dichlorobenzenes (Figure 1D). The preference in transformation was γ-HCH > α-HCH > β-HCH/δ-HCH, similar to the preference observed single isomer experiments. Therefore, *D. mccartyi* strain 195, initially spiked with γ-HCH, showed the potential to transform β-HCH and δ-HCH in addition to γ-HCH and α-HCH, although in significant lower transformation rates as α- and β-isomers were still present after 60 days but γ- and δ- isomers completely disappeared (Figure 1D). This might be due to the expression of functional genes which may boost the biodegradation potential for one isomer in the presence of other isomers as reported previously for *Pseudomonas aeruginosa* ITRC-5 under aerobic conditions (Kumar et al., 2005).

Biotransformation of HCH isomers by a γ-HCH enrichment culture

Laboratory microcosms on γ-HCH were prepared with groundwater from a mainly MCB contaminated aquifer in Bitterfeld-Wolfen (Germany), a former production site of

Lindane, shown to contain *Dehalococcoides* spp (Mészáros et al., 2013). After a lag phase of 40 days an increase in MCB concentration with a simultaneous decrease in γ-HCH was observed in all microcosms except killed controls (data not shown). After 5 months of incubation, benzene was detected as well. Microcosms amended with H_2/acetate showed highest activity as compared to bottles spiked with lactate and formate, respectively, and were therefore used for further enrichment on γ-HCH (data not shown). After the initial transformation of 25 µmol L^{-1} γ-HCH, three respective transfers with 5 % of inoculum (v/v) were prepared in fresh medium amended with γ-HCH and H_2/acetate. The third transfer was used to assess the biotransformation potential of single HCH-isomers and a mixture of them.

Initially, the biotransformation was compared for γ- and δ-HCH. While complete biotransformation of γ-HCH to MCB and benzene was observed within 120 days, only 40 % of the initial concentration of δ-HCH was transformed within the same time (Figure 2A and 2B). The preferential transformation of the γ-isomer in comparison to δ-HCH and the detected metabolites MCB and benzene were similar to *D. mccartyi* strain 195 (PCE pre-grown), however, the transformation rate in case of enrichment culture was lower as compared to *D. mccartyi* strain 195. This might be due the low cell density in case of the enrichment cultures as compared to *D. mccartyi* strain 195 cultures.

To confirm the preferential biotransformation of respective isomers, HCH isomers were added in mixture of α, β, γ and δ-isomers and compared to controls without inoculum. γ-HCH was completely transformed within 180 days whereas 33, 66 and 85 % of the initial concentration of α- , β- and δ-HCH, respectively, were still present after 180 days (Figure 2C). No transformation was observed in abiotic controls (data not shown). The detected metabolites were TeCCH, benzene and MCB and, surprisingly, traces of 2,5-dichlorophenols and trichlorobenzene. The latter metabolites were also observed in a HCH transforming methanogenic enrichment culture amended with a mixture of HCH isomers (Middeldorp et al., 1996). The preference in biotransformation of γ- and δ-HCH compared to α- and β-HCH was

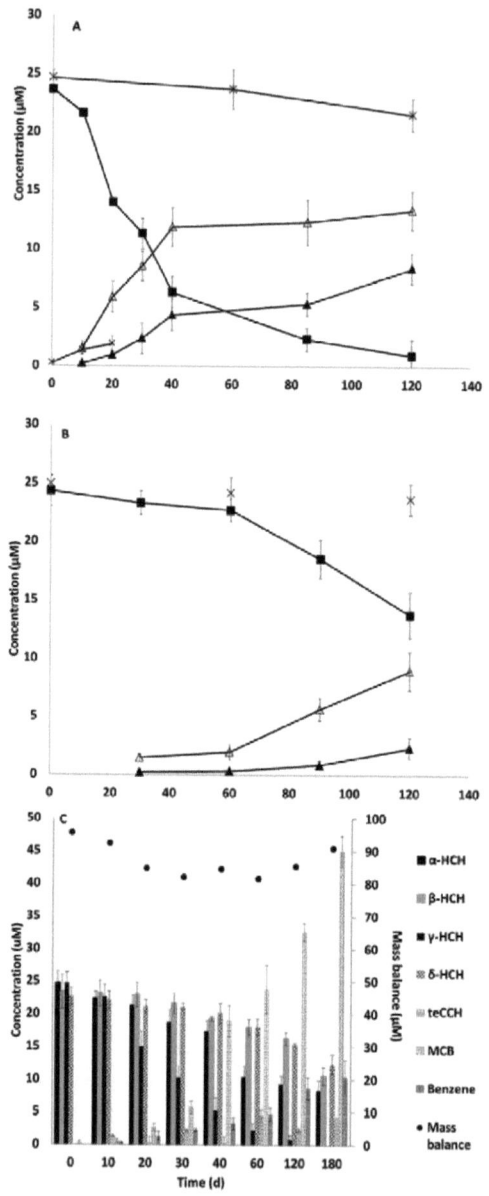

Figure 2: Transformation of γ-HCH (A), δ-HCH (B) and α-, β-, γ-, and δ- HCH isomer mixture (C) by γ-HCH degrading mixed culture. Concentrations of γ-HCH and δ-HCH (filled squares in panel A and B respectively) and the products MCB (open triangles) and benzene (filled triangles) in live cultures and controls (X). Error bars represent the standard deviation of triplicate cultures and standard deviation of triplicate analysis was less than 5 % D) Changes in concentration of the α-, β-, γ-, and δ-HCH isomers and production of intermediates by the γ-HCH enrichment culture. Error bar represent the standard deviation of duplicate experiments

identical to the HCH transformation pattern of *D. mccartyi* strain 195, however, the differences in metabolite spectrum suggest the additional presence of other enzymes and reactions in the enrichment cultures.

Although previous studies (Mészáros et al., 2013) showed the presence of *Dehalococcoides* in the groundwater used for the set-up of microcosms on γ-HCH, *Dehalococcoides* was not detected by 16S rDNA analysis in the third transfer (data not shown) used for the biotransformation experiments suggesting that other microorganisms, not detectable with the used primers, were responsible for the HCH dechlorination.

Carbon stable isotope analysis (CSIA) during biotransformation of γ-HCH

Stable carbon isotope fractionation analysis was used to evaluate the pathways involved in anaerobic γ-HCH biotransformation by *D. mccartyi* strain 195 and the γ-HCH enrichment culture. The carbon isotope fractionation by *Clostridium pasteurianum*, a strain that shown to co-metabolically transform γ-HCH to γ-tetrachlorocyclohexene as the main intermediate (Jagnow, et al., 1977), was additionally determined as a reference. Stable isotope fractionation analysis is based on the principle of favored transformation of lighter isotopes (^{12}C) during a degradation process resulting in a enrichment of heavier isotopes (^{13}C) in the residual substrate fraction and has been applied for the characterization of degradation pathways and reaction mechanisms of common groundwater contaminants such as BTEX, and MTBE (Elsner, 2010; Meckenstock et al., 2004; Vogt et al., 2008).

For all three tested anaerobic cultures, γ-HCH dechlorination was linked to significant carbon isotope fractionation (Figure 3; Table 1). A slightly stronger enrichment was observed for *D. mccartyi* strain 195 (ε_c = -5.5 ± 0.8 ‰) compared to

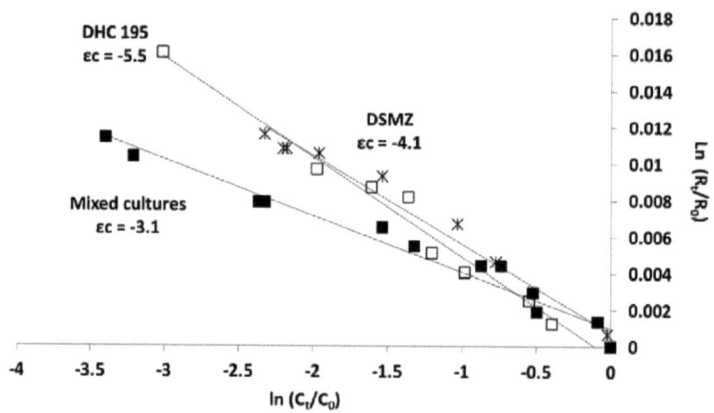

Figure 3: Rayleigh plots for D. mccartyi strain 195 (open squares), C. pasterianum DSMZ 525 (crosses) and the γ-HCH enrichment culture (filled squares).

Clostridium pasteurianum (ε_c = -4.1 ± 0.6 ‰). The range of calculated enrichment factors in our pure cultures study seemed to be higher as compares to previously reported carbon isotope enrichment factors for γ-HCH transformation by the sulfate reducers Desulfovibrio gigas (ε_c = -3.9 ± 0.6 ‰) and Desulfococcus mulitvorans (ε_c = -3.4 ± 0.5 ‰) (Badea et al., 2009). Similarities in carbon isotope fractionation suggest an analogy in the initial transformation of the HCH isomers which is confirmed by the appearance of similar metabolites in all pure culture experiments. In case of the γ-HCH enrichment culture the value of enrichment factor (ε_c = -3.1 ± 0.4 ‰) was relatively lower as compared to the pure culture experiments. The reason of lower enrichment factor might be due to the microbial competitions and several transformation processes occurring in mixed microbial communities which is also confirmed by the appearance of other metabolites such as PCCH. Due to the similarity of produced metabolites by the different tested pure cultures, independent of metabolic or co-metabolic conversion, it can be assumed that the biochemical pathway of HCH transformation was similar in the tested cultures. The data indicate that enrichment factors of anaerobic HCH dechlorination are generally relatively large, robust and can be therefore used to quantify in situ biodegradation under anoxic conditions.

Though, in comparison to the biodegradation of γ-HCH under aerobic conditions (ε_c =-1.7 ± 0.2 ‰) (Bashir et al., 2013), observed carbon enrichment factors are significantly higher and confirmed differences in metabolites and degradation pathway (Table S2, supporting information, Appendix A2). While the initial metabolite under anoxic conditions was TeCCH as result of a dichloroelemination, aerobic transformation leads to PCCH as result of dehydrochlorination (Lal et al., 2010). Additionally, enrichment factors for mixed isomer biotransformation experiments with γ-HCH enriched cultures were calculated to see if carbon stable isotope fractionation differs in single and mixed isomer experiments. The enrichment factor for γ-HCH (ε_c = -3.1 ± 0.4 ‰) was identical as calculated in single isomer biotransformation. Although complete biodegradation of other isomers was not observed, estimation enrichment factor for α-HCH (66 % degradation) and β-HCH (33 % degradation) suggested less isotope fractionation for these isomers with ε_c = -1.9 ± 0.5 ‰ and ε_c = -1.5 ± 0.2 ‰, respectively.

Conclusion

For the first time, the metabolic biotransformation of γ-HCH was shown for an isolate, *D. mccartyi* strain 195 which used hydrogen as electron donor and γ-HCH as terminal electron acceptor. *D. mccartyi* strains 195 and BTF08 as well as the enrichment culture preferentially degraded γ-HCH over the other isomers. While the *D. mccartyi* cultures preferentially degraded γ-HCH > α-HCH/δ-HCH > β-HCH, the enrichment culture preferentially transformed γ-HCH over δ-HCH > β-HCH > α-HCH. The major intermediate observed in all cultures was TCCH, with MCB and benzene as final products. Analysis of the carbon stable isotope composition during biotransformation confirmed the elucidation of degradation pathways under anoxic conditions and suggests that carbon stable isotope analysis can be used as means to quantify the *in situ* biodegradation of γ-HCH under anoxic conditions.

Table 1: Carbon enrichment factors (ε_c) for different cultures.
*Badea et al (2011) $^\$$Badea et al (2009) $^£$Bashir et al 2013

Bacteria Cultures	α-HCH	γ-HCH
	ε_c (‰)	ε_c (‰)
Anaerobic		
Dehalococcoides ethenogenes strain 195		-5.5 ± 0.8 ‰
Clostridium pasterianum	-3.7 ± 0.8 ‰ *	-4.1 ± 0.6 ‰
Enrichment culture on γ-HCH		-3.3± 0.5 ‰
Desulfovibrio gigas		-3.9 ± 0.6 ‰ $^\$$
Desulfococcus mulitvorans		-3.4 ± 0.5 ‰ $^\$$
Arobic		
Sphingobium indicum strain B90A	-1.6 ± 0.3 ‰ $^£$	-1.5 ± 0.1 ‰ $^£$
Sphingobium japonicum strain UT26	-1.0 ± 0.2 ‰ $^£$	-1.7 ± 0.2 ‰ $^£$

Acknowledgements

We thank to Ursula Gunther, Stephanie Hinke and Falk Bratfisch for the technical support. Tran Hoa Duan for helping in microscopy. Safdar Bashir was supported by University of Agriculture Faisalabad, Pakistan and the Helmholtz Impulse and Networking Fund through Helmholtz Interdisciplinary Graduate School for Environmental Research (HIGRADE).

References

Adrian L., Hansen S.K., Fung J.M., Görisch H., Zinder S.H. (2007) Growth of *Dehalococcoides* strains with chlorophenols as electron acceptors. Environmental Science & Technology, 41(7):2318-2323.

Bachmann A., De Bruin W., Jumelet J., Rijnaarts H., Zehnder A. (1988) Aerobic biomineralization of α-hexachlorocyclohexane in contaminated soil. Applied and Environmental Microbiology, 54(2):548-554.

Badea S.L., Vogt C., Weber S., Danet A.F., Richnow H.H. (2009) Stable isotope fractionation of γ-Hexachlorocyclohexane (Lindane) during reductive dechlorination by two strains of sulfate reducing bacteria. Environmental Science & Technology, 43(9):3155-3161.

Badea S.L., Vogt C., Gehre M., Fischer A., Danet A.F., Richnow H.H. (2011) Development of an enantiomer-specific stable carbon isotope analysis (ESIA) method for assessing the fate of α-hexachlorocyclohexane in the environment. Rapid Communications in Mass Spectrometry, 25(10):1363-1372.

Bashir S., Fischer A., Nijenhuis I., Richnow H.H. (2013) Enantioselective carbon stable isotope fractionation of hexachlorocyclohexane during aerobic biodegradation by *Sphingobium* spp. Environmental Science & Technology, 47(20):11432-11439.

Boyle A.W., Häggblom M.M., Young L.Y. (1999) Dehalogenation of lindane (γ-hexachlorocyclohexane) by anaerobic bacteria from marine sediments and by sulfate reducing bacteria. FEMS Microbiology Ecology, 29(4):379-387.

Buser H.R., Mueller M.D. (1995) Isomer and enantioselective degradation of hexachlorocyclohexane isomers in sewage sludge under anaerobic conditions. Environmental Science & Technology, 29(3):664-672.

Cichocka D., Imfeld G., Richnow H.H., Nijenhuis I. (2008) Variability in microbial carbon isotope fractionation of tetra-and trichloroethene upon reductive dechlorination. Chemosphere, 71(4):639-648.

Doesburg W., Eekert M.H., Middeldorp P.J., Balk M., Schraa G., Stams A.J. (2005) Reductive dechlorination of β-hexachlorocyclohexane (β-HCH) by a *Dehalobacter* species in coculture with a *Sedimentibacter* sp. FEMS Microbiology Ecology, 54(1):87-95.

Elango V., Cashwell J.M., Bellotti M.J., Marotte R., Freedman D.L. (2010) Bioremediation of hexachlorocyclohexane isomers, chlorinated benzenes, and chlorinated ethenes in soil and fractured dolomite. Bioremediation Journal, 14(1):10-27.

Elsner M. (2010) Stable isotope fractionation to investigate natural transformation mechanisms of organic contaminants: principles, prospects and limitations. Journal of Environmental Monitoring, 12(11):2005-2031.

Hendrickson E.R., Payne J.A., Young R.M., Starr M.G., Perry M.P., Fahnestock S., Ellis D.E., Ebersole R.C. (2002) Molecular analysis of *Dehalococcoides* 16S ribosomal DNA from chloroethene-contaminated sites throughout North America and Europe. Applied and Environmental Microbiology, 68(2):485-495.

Heritage A., MacRae I. (1977) Degradation of lindane by cell-free preparations of *Clostridium sphenoides*. Applied and Environmental Microbiology, 34(2):222-224.

Jagnow G., Haider K., Ellwardt P.C.H.R. (1977a) Anaerobic dechlorination and degradation of hexachlorocyclohexane isomers by anaerobic and facultative anaerobic bacteria. Archives of Microbiology, 115(3):285-292.

Kaufhold T., Schmidt M., Cichocka D., Nikolausz M., Nijenhuis I. (2013) Dehalogenation of diverse halogenated substrates by a highly enriched *Dehalococcoides* containing culture derived from the contaminated mega- site in Bitterfeld. FEMS Microbiology Ecology, 83(1):176-188.

Krajmalnik-Brown R., Sung Y., Ritalahti K.M., Saunders F.M., Loffler F.E. (2007) Environmental distribution of the trichloroethene reductive dehalogenase gene (tceA) suggests lateral gene transfer among *Dehalococcoides*. FEMS Microbiology Ecology, 59(1):206-214.

Kumar M., Chaudhary P., Dwivedi M., Kumar R., Paul D., Jain R.K., Garg S.K., Kumar A. (2005) Enhanced biodegradation of β-and δ-hexachlorocyclohexane in the presence of α-and γ-isomers in contaminated soils. Environmental Science & Technology, 39(11):4005-4011.

Lal R., Pandey G., Sharma P., Kumari K., Malhotra S., Pandey R., Raina V., Kohler H.P.E., Holliger C., Jackson C., Oakeshott J.G. (2010) Biochemistry of Microbial Degradation of hexachlorocyclohexane and prospects for bioremediation. Microbiology and Molecular Biology Reviews, 74(1):58-80.

Löffler, F. E., Ritalahti, K. M., & Zinder, S. H. (2013). Dehalococcoides and reductive dechlorination of chlorinated solvents. In Bioaugmentation for groundwater remediation (pp. 39-88). Springer New York.

MacRae I., Raghu K., Bautista E. (1969) Anaerobic degradation of the insecticide lindane by *Clostridium* sp. Nature, 221:859-860

Mariotti A., Germon J., Hubert P., Kaiser P., Letolle R., Tardieux A., Tardieux P. (1981) Experimental determination of nitrogen kinetic isotope fractionation: some principles; illustration for the denitrification and nitrification processes. Plant and Soil, 62(3):413-430.

Maymo-Gatell X. (1997). *Dehalococcoides ethenogenes* Strain 195, A novel eubacterium that reductively dechlorinates tetrachloroethene (PCE) to ethene. Cornell Univ Ithaca NY Dept. of Civil and Environmental Engineering.

Meckenstock R.U., Morasch B., Griebler C., Richnow H.H. (2004) Stable isotope fractionation analysis as a tool to monitor biodegradation in contaminated acquifers. Journal of Contaminant Hydrology, 75(3-4):215-255.

Mehboob F., Langenhoff A.A., Schraa G., Stams A.J. (2013) Anaerobic Degradation of Lindane and Other HCH Isomers, Management of microbial resources in the environment, Springer. pp. 495-521.

Mészáros É., Imfeld G., Nikolausz M., Nijenhuis I. (2013) Occurrence of *Dehalococcoides* and reductive dehalogenase genes in microcosms, a constructed wetland and groundwater from a chlorinated ethene contaminated field site as indicators for in situ reductive dehalogenation. Water, Air, & Soil Pollution, 224(11):1-12.

Middeldorp P.J.M., Jaspers M., Zehnder A.J.B., Schraa G. (1996) Biotransformation of α-, β-, γ-, and δ-hexachlorocyclohexane under methanogenic conditions. Environmental Science & Technology, 30(7):2345-2349.

Nagasawa S., Kikuchi R., Nagata Y., Takagi M., Matsuo M. (1993) Aerobic mineralization of γ-HCH by *Pseudomonas paucimobilis* UT26. Chemosphere, 26(9):1719-1728.

Nijenhuis I., Nikolausz M., Köth A., Felföldi T., Weiss H., Drangmeister J., Großmann J., Kästner M., Richnow H.-H. (2007) Assessment of the natural attenuation of chlorinated ethenes in an anaerobic contaminated aquifer in the Bitterfeld/Wolfen area using stable isotope techniques, microcosm studies and molecular biomarkers. Chemosphere, 67(2):300-311.

Phillips T.M., Seech A.G., Lee H., Trevors J.T. (2005) Biodegradation of hexachlorocyclohexane (HCH) by microorganisms. Biodegradation, 16(4):363-392.

Popp P., Brüggemann L., Keil P., Thuß U., Weiß H. (2000) Chlorobenzenes and hexachlorocyclohexanes (HCHs) in the atmosphere of Bitterfeld and Leipzig (Germany). Chemosphere, 41(6):849-855.

Pöritz M., Goris T., Wubet T., Tarkka M.T., Buscot F., Nijenhuis I., Lechner U., Adrian L. (2013) Genome sequences of two dehalogenation specialists-*Dehalococcoides mccartyi* strains BTF08 and DCMB5 enriched from the highly polluted Bitterfeld region. FEMS Microbiology Letters, 343(2):101-104

Ricking M., Schwarzbauer J. (2008) HCH residues in point-source contaminated samples of the Teltow Canal in Berlin, Germany. Environmental Chemistry Letters, 6:83-89.

Sahu S.K., Patnaik K., Bhuyan S., Sreedharan B., Kurihara N., Adhya T., Sethunathan N. (1995) Mineralization of. α.-,. γ.-, and. β.-Isomers of Hexachlorocyclohexane by a soil bacterium under aerobic conditions. Journal of Agricultural and Food Chemistry, 43(3):833-837.

Schmidt M., Wolfram D., Birkigt J., Ahlheim J., Paschke H., Richnow H.-H., Nijenhuis I. (2014) Iron oxides stimulate microbial monochlorobenzene in situ transformation in constructed wetlands and laboratory systems. Science of the Total Environment, 472:185-193.

Seshadri R., Adrian L., Fouts D.E., Eisen J.A., Phillippy A.M., Methe B.A., Ward N.L., Nelson W.C., Deboy R.T., Khouri H.M. (2005) Genome sequence of the PCE-dechlorinating bacterium *Dehalococcoides ethenogenes*. Science, 307(5706):105-108.

Van Eekert M.H., Van Ras N.J., Mentink G.H., Rijnaarts H.H., Stams A.J., Field J.A., Schraa G. (1998) Anaerobic transformation of β-hexachlorocyclohexane by methanogenic granular sludge and soil microflora. Environmental Science & Technology, 32(21):3299-3304.

Vijgen J., Abhilash P., Li Y.F., Lal R., Forter M., Torres J., Singh N., Yunus M., Tian C., Schäffer A. (2011) Hexachlorocyclohexane (HCH) as new Stockholm convention POPs - a global perspective on the management of Lindane and its waste isomers. Environmental Science and Pollution Research, 18(2):152-162.

Vogt C., Cyrus E., Herklotz I., Schlosser D., Bahr A., Herrmann S., Richnow H.-H., Fischer A. (2008) Evaluation of toluene degradation pathways by two-dimensional stable isotope fractionation. Environmental Science & Technology, 42(21):7793-7800.

Walker K., Vallero D.A., Lewis R.G. (1999) Factors influencing the distribution of lindane and other hexachlorocyclohexanes in the environment. Environmental Science & Technology, 33(24):4373-4378.

Willett K.L., Ulrich E.M., Hites R.A. (1998) Differential toxicity and environmental fates of hexachlorocyclohexane isomers. Environmental Science & Technology, 32(15):2197-2207.

Zinder S.H. (1998) Methanogens. In: Burlage, RS (Ed.) Techniques in Microbial Ecology. Oxford University Press, Oxford, pp 113-136

3.3

The Application of Compound Specific Isotope Analysis to Characterize Abiotic Reaction Mechanisms of α-Hexachlorocyclohexane

Ning Zhang[1], Safdar Bashir[1], Jinyi Qin[2], Anko Fischer[1,3], Ivonne Nijenhuis[1], Hartmut Herrmann[4], Lukas Y. Wick[2], and Hans-Hermann Richnow[1]

[1]Department of Isotope Biogeochemistry, Helmholtz Centre for Environmental Research – UFZ, Permoserstraße 15, 04318, Leipzig, Germany

[2]Department of Environmental Microbiology, Helmholtz Centre for Environmental Research-UFZ, Permoserstraße 15, 04318 Leipzig, Germany

[3]Isodetect - Company for Isotope Monitoring, Permoserstr. 15, 04318 Leipzig, Germany

[4]Department of Chemistry, Leibniz Institute for Tropospheric Research (TROPOS), Chemistry Dept., Permoserstraße 15, 04318 Leipzig, Germany

Abstract

A systematic investigation of environmentally relevant transformation processes of alpha-hexachlorocyclohexane (α-HCH) was performed in order to explore the potential of compound-specific stable isotope analysis (CSIA) to characterize reaction mechanisms. The carbon isotope enrichment factors (ε_c) for the chemical transformations of α-HCH via direct photolysis, indirect photolysis (UV/H_2O_2), hydrolysis, electro-reduction or reduction by Fe^0 nanoparticles were quantified and compared to those previously published for biodegradation. Hydrogen abstraction by hydroxyl radicals generated by UV/H_2O_2 led to ε_c of −1.9 ± 0.2 ‰ with an apparent kinetic carbon isotope effect ($AKIE_C$) of 1.012 ± 0.001. Dehydrochlorination by alkaline hydrolysis yielded ε_c of −7.6 ± 0.4 ‰ with $AKIE_c$ of 1.048 ± 0.003. Dechlorination *either by homolytic bond cleavage* in direct photolysis (ε_c = −2.8 ± 0.2 ‰) or single-electron transfer in electro-reduction (ε_c = −3.8 ± 0.4 ‰) corresponded to $AKIE_c$ of 1.017 ± 0.001 and 1.023 ± 0.003 respectively. Dichloroelimination catalyzed by Fe^0 via two-electron transfers resulted in ε_c of −4.9 ± 0.1 ‰. $AKIE_c$ values assuming either a concerted or a stepwise mechanism were 1.030 ± 0.0006 and 1.015 ± 0.0003 respectively. Contrary to biodegradation, no enantioselectivity of α-HCH was observed in chemical reactions, which might be used to identify chemical and biological *in situ* transformations.

3.3.1 Introduction

Hexachlorocyclohexane (HCH) isomers have been listed in the Stockholm Convention on Persistent Organic Pollutants (POPs) (Lallas, 2001), that aims to eliminate or restrict the production and use of POPs. Although production of HCH is mostly banned in countries signing the Stockholm Convention, residues from previous application, improper waste disposal and storage require long-term monitoring in order to assess sources and sinks. HCH isomers are ubiquitous in the environment due to their previously widespread global usage and physicochemical recalcitrance toward decomposition (Abhilash and Singh, 2009; Jiang et al., 2009). Technical-grade HCH, an important insecticide formulation in agriculture, forestry, and wood preservative, consists mainly of five major isomers: α (60-70%), β (5-12%), γ (10-12%), δ (6-10%), and ε (3-4%) (Vijgen, 2006). The intensive use of technical-grade HCH has released large amounts of α-HCH into the environment and its persistence led to the presence in environmental and biological samples until today (Li, 1999). Therefore, understanding of the fate of α-HCH is of great importance for long-term cleanup activities. Although HCH isomers are highly resistant to degradation under the conditions prevailing in the environment, biological and chemical transformation processes and related remediation strategies have been explored to clean up HCH contaminated sites (Lal et al., 2010; Ruiz et al., 2012; Singh et al., 2013).

An attempt on the characterization and quantification of degradation is a prerequisite for evaluating the environmental fate of HCH isomers and assessing the efficiency of remediation techniques. Aerobic and anaerobic microbial degradation of HCH isomers (Bhatt et al., 2007; Murthy and Manonmani, 2007) likewise require evaluation for bioremediation at contaminated sites. In addition to biotransformation, radical oxidation by photo-Fenton process (Nitoi et al., 2013) or by photocatalytic degradation (Zaleska et al., 2000) has been discussed as a strategy for treating γ-HCH-polluted water. Electrochemical reduction of γ-HCH applying a modified $NiCo_2O_4$ electrode (Srivastava, 2006) may be used as a remediation technology in the future. Nanoscale zero-valent iron (Fe^0) has been discussed for transformation of HCH isomers in contaminated groundwater and soil representing a new generation of environmental remediation technology (Elliott et al., 2008; Elliott et al., 2009; Fu et al., 2014; Singh et al., 2011; Wang et al., 2009; Zhang, 2003).

Transformation of HCH is typically initiated by hydrogen abstraction, dehydrochlorination, dechlorination or dichloroelimination. Hydrogen abstraction can be found in radical oxidation processes via reactive hydroxyl radicals (OH) formed by photo-induced H_2O_2 decomposition (Nienow et al., 2008) or by TiO_2-enhanced photocatalysis (Zaleska et al., 2000). Dehydrochlorination takes place during hydrolysis of γ-HCH under alkaline conditions (Liu et al., 2003), which might be a similar mechanism to aerobic degradation of α-HCH by *Sphingomonas paucimobilis* B90A (Suar et al., 2005b). Dechlorination initiated by direct photolysis of short-wavelength UV irradiation (Hamada et al., 1982) or by single-electron transfer via electrodes (Srivastava, 2006) may directly transform HCH isomers into a pentachlorocyclohexyl radical by cleavage of a C−Cl bond. Two-electron transfers to the molecule initiate dichloroelimination in a stepwise or a concerted mode (Fletcher et al., 2009; Tobiszewski and Namiesnik, 2012). Transformation of HCH by dichloroelimination can be induced by Fe^0 nanoparticles (Elliott et al., 2008; Elliott et al., 2009; Singh et al., 2011). Similarly, anaerobic degradation of α-HCH by *Clostridium pasterianum* proceeds via dichloroelimination whereby two C–Cl bonds are reductively cleaved leading to 3,4,5,6-tetrachlorocyclohexene (TeCCH) (Badea et al., 2011).

Information on the transformation mechanism is essential for assessing the fate of HCH isomers in the environment. However, it is difficult to distinguish among different reaction mechanisms solely based on analysis of the reaction intermediates. For instance, the first step in aerobic degradation of α-HCH is dehydrochlorination, a similar mechanism compared to alkaline hydrolysis. The pattern of intermediates alone often does not allow characterizing specific degradation pathways.

Carbon stable isotope analysis (CSIA) is a promising tool for characterizing transformation pathways of pollutants in the environment (Aelion et al., 2009). The stable isotope fractionation identified by CSIA mainly depends on kinetic isotope effect (KIE) associated with the cleavage of chemical bonds, which is specific for a reaction mechanism. CSIA has been successfully applied to aerobic and anaerobic biotransformation of α- and γ-HCH and distinct isotope fractionation patterns for carbon were observed (Badea et al., 2009; Badea et al., 2011; Bashir et al., 2013). α-HCH exists in two enantiomeric forms. The carbon isotope ratios of the α-HCH enantiomers were determined suggesting that enantiomer-specific stable isotope analysis (ESIA) can be used as a complementary approach to CSIA for assessing biodegradation of α-HCH in the environment (Bashir et al., 2013). However, although CSIA for HCH biodegradation has been a subject of current studies, information regarding isotope fractionation upon chemical degradation of HCH is rare. There is only a single study on carbon isotope fractionation for the hydrolysis of α- and γ-HCH (Peng et al., 2004).

Due to the limited knowledge on KIEs for chemical degradation of α-HCH, we determined carbon isotope fractionation for environmentally relevant transformation processes of α-HCH in order to explore the potential of CSIA and ESIA for the identification of reaction mechanisms. Different mechanisms were investigated (Scheme 1) for chemical reactions including direct photolysis, photochemical oxidation by OH radicals, alkaline hydrolysis, electrochemical reduction and reduction by Fe^0 nanoparticles. Apparent kinetic carbon isotope effects ($AKIE_c$) were calculated and compared with putative KIEs for the elucidation and distinction of α-HCH reaction mechanisms. Studies on the enantioselectivity of α-HCH were also performed to evaluate if various chemical reactions can be distinguished from biotransformation.

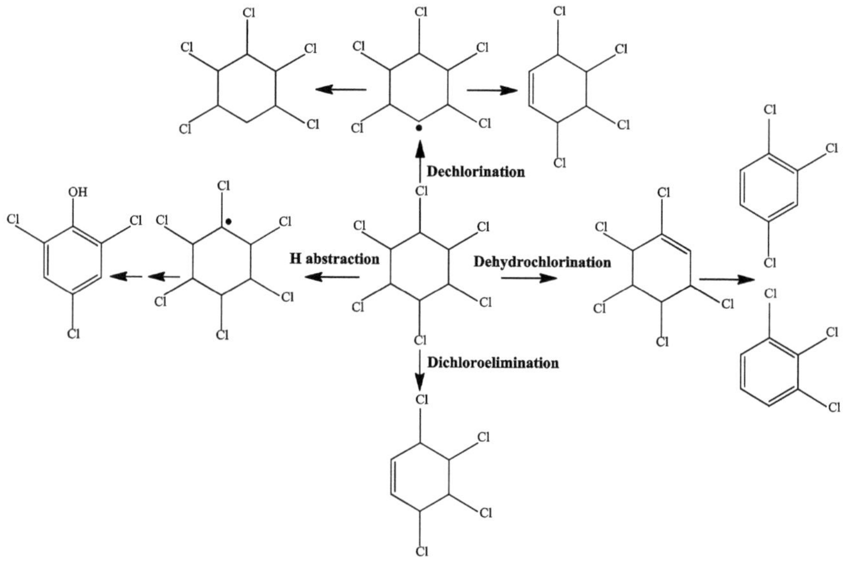

Scheme 1: Transformation pathways of α-HCH in biological and chemical processes.

3.3.2 Materials and methods

Information on chemicals used in this study and on the preparation of α-HCH stock solution is provided in the Supporting Information (SI).

Direct and indirect photolysis

The photochemical reactor system consisted of a 200 mL Pyrex cylindrical flask with a quartz window whose surface area was approximately 28 cm^2 (Figure S1, supporting information, Appendix A3). Irradiation was achieved by applying a 150 W xenon lamp (Type No.: L2175, Hamamatsu, Japan), which covered a broad continuous spectrum from 185 nm to 2000 nm. Direct photolysis and indirect photolysis were conducted in a temperature-controlled water system at 20°C at two wavelength ranges: ≥ 185 nm and ≥ 280 nm, respectively. A filter with a 280 nm cut-off wavelength was applied, because range of larger wavelengths is typical at the Earth's surface. The distance between the reactor and the light source was 10 cm. The reaction solution was magnetically stirred during the irradiation experiments.

For direct photolysis at $\lambda \geq 185$ nm, the light source was directly faced to the quartz window of the reactor without the filter with a 280 nm cut-off wavelength. The α-HCH stock solution (1 mg L^{-1}) was bubbled with argon for 1 hour to remove oxygen before turning on the lamp. For indirect photolysis, the filter with a 280 nm cut-off wavelength was used and hydrogen peroxide was added to the α-HCH stock solution (molar ratio of H$_2$O$_2$ to α-HCH was 100:1). The solution was mixed with a magnetic stirrer during the whole experiment. For sampling at different intervals, the reaction mixture was transferred through a rubber septum of the reactor and filled into gas-tight vials using a syringe. *n*-Pentane (0.5 mL) containing hexachlorobenzene (HCB, 1 mg L^{-1}, as internal standard) was added to the vials for the extraction of α-HCH and its degradation products. The mixture was shaken for at least one hour prior to the phase separation. To avoid evaporation, the *n*-pentane phase was transferred at 10°C to a vial and the vials were stored at -20°C until the analysis. Each experiment in this study was conducted with at least three repetitions.

Alkaline hydrolysis

A volume of 300 mL Na$_2$CO$_3$-NaHCO$_3$ buffer solution (pH 9.78, 10 mM) was spiked with α-HCH (1000 mg L^{-1}) dissolved in acetone achieving a concentration of 1 mg L^{-1} in the buffer solution. Twenty of 20 mL vials were filled with 15 mL of this buffer solution for each of them and closed with Teflon coated butyl rubber septa and crimped. All solutions were put on a shaker for reaction at 30°C until sampling. At different intervals, n-pentane (0.5 mL) with HCB (1 mg L^{-1}) was added to extract α-HCH and its degradation products from the water phase.

Electrochemical reduction

For electrochemical reduction of α-HCH, two inert Ti/Ir coated grids (13 cm × 3.2 cm) were inserted as electrodes into a stirred 500 mL beaker, which was filled with 300 mL of the α-HCH stock solution (1 mg L^{-1}) and covered with a Teflon membrane to prevent evaporation (Figure S2, supporting information, Appendix A3). The reaction was performed at 25°C and using an electric field strength of 6.3 V cm^{-1} (1.68 mA cm^{-2}). The experiment in the absence of an electric field served as a control. At given times, samples were taken by a syringe. α-HCH and its degradation products were extracted with n-pentane (0.5 mL) containing HCB (1 mg L^{-1}).

Reduction by Fe^0 nanoparticles

Fe^0 nanoparticles were synthesized using a method in which ferrous sulfate (0.25 M) was reduced by sodium borohydride (0.5 M) in aqueous solution as described elsewhere (Wang and Zhang, 1997; Zhang et al., 1998). The synthesized Fe^0 nanoparticles were washed with ethanol, purged with nitrogen for drying, and refrigerated in a sealed polyethylene container under absolute ethanol until use. The reaction of α-HCH with synthesized Fe^0 nanoparticles was performed in a 250 mL serum flask. Fe^0 nanoparticles (0.3 g) were added to 200 mL of the α-HCH stock solution (1 mg L^{-1}) and the flask was subsequently capped with a Teflon valve. The reaction mixture was mixed with a magnetic stirrer during the whole experiment at room temperature. Periodically, 15 mL of the aqueous solution was taken by a syringe and passed through a 0.20 μm syringe filter. Then α-HCH and its degradation products were extracted with n-pentane (0.5 mL) containing HCB (1 mg L^{-1}).

Analytical methods

The concentration of α-HCH was determined by gas chromatograph equipped with a flame ionization detector (GC-FID) and degradation products were analyzed by gas chromatograph-mass spectrometry (GC-MS) as described elsewhere (Bashir et al., 2013). Carbon stable isotope ratios of α-HCH and its enantiomers were measured by gas chromatograph-combustion-isotope ratio mass spectrometry (GC-C-IRMS) (Bashir et al., 2013). A detailed description of analytical methods is provided in SI.

3.3.3 Results and discussion

Direct and indirect photolysis

The stock solution of α-HCH showed a maximum UV absorption at λ = 252 nm (Figure S3, supporting information, Appendix A3), which was in agreement with the reported high absorption at λ = 255 nm (Fiedler et al., 1993). The carbon isotope ratio of α-HCH showed ^{13}C enrichment from −27.8 ± 0.3 ‰ to −21.0 ± 0.3 ‰ after 91 % removal of α-HCH (Figure 1A). The Rayleigh equation was applied as described in SI in order to determine carbon isotope enrichment factors (ε_c). Based on this approach, ε_c of −2.8 ± 0.2 ‰ was determined for direct photolysis of α-HCH. Indirect photolysis at a wavelength ≥ 280 nm by a UV/H_2O_2 oxidation process resulted in 85 % degradation within 12 hours concomitant with a ^{13}C enrichment from −27.7 ± 0.1 ‰ to −24.2 ± 0.1 ‰ (Figure 1A), yielding ε_c of −1.9 ± 0.2 ‰. As indicated by Figure 1B, direct photolysis (≥ 185 nm) and indirect photolysis (≥ 280 nm, UV/H_2O_2) followed pseudo first order kinetics (R^2 = 0.99 and 0.99, respectively) with rate constants of 0.34 h^{-1} and 0.16 h^{-1}, respectively. Control experiments either with H_2O_2 in the absence of light or only with UV irradiation (≥ 280 nm) showed almost stable α-HCH concentrations (Figure S4, supporting information, Appendix A3). It indicated that the sole presence of H_2O_2 or longer wavelength ≥ 280 nm did not induce the transformation of α-HCH under the experimental conditions used.

1,2,3,4,5-pentachlorocyclohexane ($C_6H_7Cl_5$, PCCHa) and 3,4,5,6-tetrachlorcyclohexene (TeCCH) were determined as major transformation products for direct photolysis (≥ 185 nm) of α-HCH (Figure S5, supporting information, Appendix A3). It was supposed that UV irradiation initiated homolytic bond cleavage of a C–Cl bond leading to the formation of a pentachlorocyclohexyl radical and a chlorine radical. The pentachlorocyclohexyl radical can abstract a hydrogen atom from another α-HCH to form PCCHa. There were two possible pathways for the following formation of TeCCH: i) a radical disproportionation process, in which the chlorine radical can abstract a chlorine atom from the pentachlorocyclohexyl radical leading to a double bond; ii) a second C–Cl bond cleavage induced by direct photolysis of the pentachlorocyclohexyl radical (Scheme S1, supporting information, Appendix A3).

Indirect photolysis of α-HCH was performed by dissociating H_2O_2 via UV irradiation (≥ 280 nm) to form OH radicals resulting in the formation of 2,4,6-trichlorophenol as the main transformation product (Figure S6, supporting information, Appendix A3). The initial step of indirect photolysis was the abstraction of hydrogen by OH radicals leading to the formation of an α-HCH radical (Wang et al., 2009). The second step might be a simultaneous multi-dehydrochlorination to remove three sets of chlorine and hydrogen atoms concomitant with the formation of a 1,3,5-trichlorobenzene radical, which can be stabilized by hydrogen abstraction from another α-HCH. The further addition of an OH radical to the aromatic ring and the subsequent reaction with oxygen led to the formation of a peroxyl radical, which was then stabilized to form 2,4,6-trichlorophenol via hydroperoxyl radical elimination (Scheme S2, supporting information, Appendix A3) (Von Sonntag and Schuchmann, 1991).

Alkaline hydrolysis

Alkaline hydrolysis caused 87 % decrease in α-HCH concentration within 13 days (degradation). The carbon isotope ratio of α-HCH exhibited a clearly high ^{13}C enrichment from −27.8 ± 0.02 ‰ to −12.9 ± 0.4 ‰ (Figure 1C) resulting in ε_c of −7.6 ± 0.4 ‰. So far, the only reported carbon isotope enrichment factor for alkaline hydrolysis of α-HCH was determined as −8.5 ‰ (Peng et al., 2004), which was similar to the value obtained in our study. The reaction followed pseudo first order kinetic (R^2=0.99) (Figure 1D) with a rate constant of 0.0064 h^{-1} (11×10^{-5} min^{-1}), which was in agreement with a previous reported rate constant of 9.74×10^{-5} min^{-1} obtained at pH 9 and 30°C (Ngabe et al., 1993).

The main transformation products for alkaline hydrolysis of α-HCH were 1,3,4,5,6-pentachlorocyclohexene ($C_6H_5Cl_5$, PCCH), 1,2,4-trichlorobenzene (TCB) and 1,2,3-TCB (Figure S7, supporting information, Appendix A3), indicating a dehydrochlorination mechanism similar to aerobic degradation (Suar et al., 2005b). The initial step of hydrolysis should be the formation of PCCH through a concerted bimolecular elimination mechanism (E2 reaction) in which the C–H bond cleavage occurred simultaneously with the C–Cl bond cleavage leading to the formation of a double bond (Liu et al., 2003). In subsequent hydrolysis steps, TCB isomers may be formed due to simultaneous eliminations of two chlorine and two hydrogen atoms from PCCH (Scheme S3, supporting information, Appendix A3).

Electrochemical reduction

For electrochemically induced transformation at 80 % degradation within 10 hours, the carbon isotope ratio of α-HCH exhibited a ^{13}C enrichment from −27.8 ± 0.1 ‰ to −20.0 ± 0.1 ‰ (Figure 1E) which resulted in ε_c of −3.8 ± 0.4 ‰. The reaction followed pseudo first order kinetic (R^2=0.99) with a rate constant of 0.15 h^{-1} (Figure 1F). Control experiment without electrodes showed insignificant losses of α-HCH (Figure S8). GC-MS spectra indicated the same products for electrochemically induced transformation as for direct photolysis (i.e. PCCHa and TeCCH) (Figure S9, supporting information, Appendix A3) but a different transformation pathway was proposed (Scheme S4). A single-electron transfer mechanism might be expected which was in accordance with previous studies indicating an electron transfer mechanism for electrochemical reduction (Evans, 2008). It was assumed that single-electron transfer of α-HCH led to a C−Cl bond cleavage generating a pentachlorocyclohexyl radical. The hydrogen atom from the electrolysis of H_2O can quench the pentachlorocyclohexyl radical to produce PCCHa. The formation of TeCCH was possibly a second electron transfer to the pentachlorocyclohexyl radical.

Reduction by Fe⁰ nanoparticles

Reduction by Fe^0 nanoparticles led to 90 % degradation of α-HCH within 2.5 hours (Figure 1E) and yielded the highest rate constant (0.86 h^{-1}) of all the reactions conducted in this study (Figure 1F). The carbon isotope ratio of α-HCH was enriched in ^{13}C from −27.9 ± 0.1 ‰ to −16.8 ± 0.1 ‰ (Figure 1E) and gave ε_c of −4.9 ± 0.1 ‰. TeCCH was the only product identified by GC-MS (Figure S10, supporting information, Appendix A3). Therefore, electrons provided by Fe^0 may attack α-HCH by two-electron transfers eliminating two chlorine atoms from the ring and forming a double bond (Scheme S5, supporting information, Appendix A3). However, it was unknown whether the reaction proceeded stepwise (i.e., one C−Cl bond was broken in the transition state and the rate limiting step involved only one carbon atom), or via a concerted mode (i.e., two C–Cl bonds were broken simultaneously and the rate limiting step involved two carbon atoms) (Elsner et al., 2007).

Enantiomer-specific stable isotope analysis (ESIA)

It is known that chemical transformation processes do not exhibit enantioselectivity (Bidleman et al., 2012; Hühnerfuss et al., 1993). Due to the same reactivity of enantiomers, it can be expected that chemical reactions would lead to similar isotope fractionation of enantiomers. In order to confirm this assumption, carbon isotope fractionation of individual enantiomers of α-HCH was investigated for direct photolysis, indirect photolysis through UV/H_2O_2, hydrolysis, electro-reduction or reduction by Fe^0 nanoparticles. As expected, no enantioselective transformation of α-HCH was observed (Figure S11, supporting information, Appendix A3). In contrast to enantioselective biodegradation, the carbon isotope enrichment factors of individual enantiomers ($\varepsilon_c(+)$, $\varepsilon_c(-)$) upon chemical transformation were statistically identical with each other and with

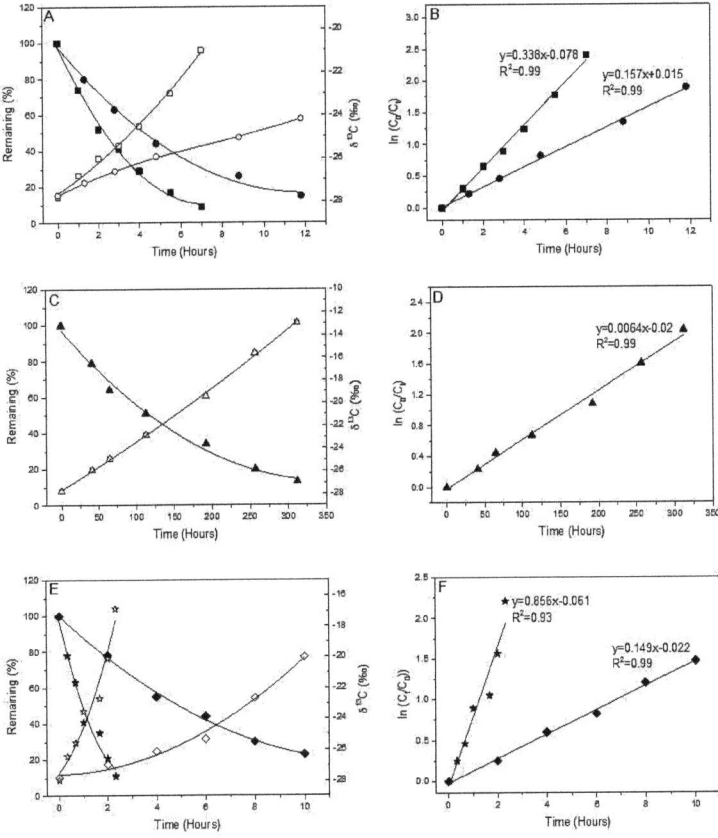

Figure 1: Remaining fraction (closed symbols) and carbon isotope ratios (open symbols) of α-HCH for transformation by (A) direct photolysis (squares) and UV/H_2O_2 process (circles), (C) alkaline hydrolysis (triangles), (E) electrochemical reduction (rhombuses) and reduction by Fe^0 (stars); Pseudo first order kinetics (closed symbols) for (B) direct photolysis (squares) and UV/H_2O_2 process (circles), (D) alkaline hydrolysis (triangles), (F) electrochemical reduction (rhombuses) and reduction by Fe^0 (stars).

the respective ε_c of bulk α-HCH (Table 1). Thus, the difference in carbon isotope fractionation between enantiomers can provide a direct evidence for α-HCH biodegradation.

3.3.4 Discussion

Carbon isotope fractionation of the reaction mechanisms

Significant differences in carbon isotope fractionation of *α*-HCH were observed for various chemical transformation processes (Figure 2 and Scheme 1). The reaction with OH radicals formed by photo-induced dissociation of H_2O_2 exhibited the smallest carbon isotope fractionation (ε_c of −1.9 ± 0.2 ‰). In this reaction, the rate limiting step was supposed to be H abstraction via cleavage of a C−H bond. Direct photolysis and electro-reduction resulted in moderate carbon isotope fractionation (ε_c of −2.8 ± 0.2 ‰ and −3.8 ± 0.4 ‰, respectively). The rate limiting step was considered to be dechlorination via cleavage of a C−Cl bond either by homolytic bond cleavage or single-electron transfer. Compared to direct photolysis and electro-reduction, reduction by Fe^0 nanoparticles led to a slightly higher carbon isotope fractionation (ε_c of −4.9 ± 0.1 ‰). For this process, dichloroelimination involving two-electron transfers to *α*-HCH with the cleavage of two C−Cl bonds could be expected as the initial step of the reaction mechanism. It is known that the cleavage of a bond formed by a heavier element tends to induce a larger isotope effect compared to a bond formed by a lighter element (Aelion et al., 2009). Hence, larger carbon isotope fractionations were observed for C−Cl bond cleavage reactions (direct photolysis, electro-reduction, reduction by Fe^0) compared to C−H bond cleavage by OH radical-induced oxidation (indirect photolysis). The largest carbon isotope fractionation (ε_c of −7.6 ± 0.4 ‰) was found for alkaline hydrolysis by a dehydrochlorination mechanism with simultaneous cleavage of a C−H bond and a C−Cl bond. Hypothetically, the reason for high carbon isotope fractionation may be that two carbon atoms were involved simultaneously leading to a complex transition state and providing an advantage for the reaction kinetic of lighter isotopologues.

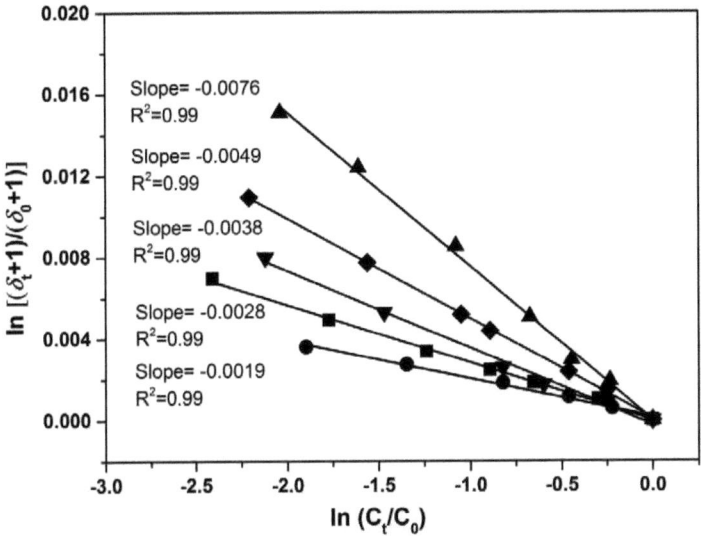

Figure 2: Double logarithmic plot according to the Rayleigh equation (eq. S3, supporting information, Appendix A3) for α-HCH transformation by direct photolysis (squares), indirect photolysis under UV/H$_2$O$_2$ (circles), alkaline hydrolysis (upward triangles), electrochemical reduction (downward triangles) and reduction by Fe^0 (rhombuses).

Comparison of chemical and biological isotope fractionation

Dehydrochlorination of α-HCH by alkaline hydrolysis and aerobic biodegradation yielded PCCH and TCBs as products indicating an E2 elimination reaction for both transformation pathways. In comparison to alkaline hydrolysis (ε_c of −7.6 ± 0.4 ‰), aerobic biodegradation (ε_c of −1.6 ± 0.3 ‰ and −1.0 ± 0.2 ‰ in experiments with *Sphingobium indicum* strain B90A and *Sphingobium japonicum* strain UT26, respectively (Bashir et al., 2013)) showed much lower carbon isotope fractionations. It might suggest different mechanisms and rate limitations in the transition stage of bond cleavage. The binding of the substrate to the enzyme during aerobic biodegradation may contribute to the lower carbon isotope fractionation as compared to direct bond cleavage in pure chemical reactions (Nijenhuis et al., 2005). Enzyme binding might be a rate limiting step prior to the isotope sensitive bond cleavage in biological systems (Northrop, 1981), complicating the mechanistic interpretation of kinetic isotope fractionation pattern. This rate limitation can mask the extent of isotope fractionation for biotransformation.

Further information on rate limitation associated with enzyme-substrate binding can be obtained when comparing the isotope fractionation between α-HCH enantiomers. For *Sphingobium japonicum* strain UT26, the enzyme LinA (HCH dehydrochlorinase) is thought to catalyze dehydrochlorination of both enantiomers by the same catalytic mechanism (Lal et al., 2010). The difference in isotope fractionation may thus be related to rate limitation induced by enzyme-substrate binding. Assuming that the catalytic difference in ε_c of the α-HCH enantiomers ($\Delta\varepsilon_c = 1.8$ ‰, Table 1) quantitatively demonstrated the masking effect on carbon isotope fractionation by binding to enzyme. A similar difference in ε_c of the enantiomers ($\Delta\varepsilon_c = 1.4$ ‰) was observed for Sphingobium indicum strain B90A degrading α-HCH with the same enzyme LinA, confirming that binding of enantiomers to the enzyme played a role in the overall process of isotope fractionation. Taking into account that isotope fractionation of the (+)-enantiomer was less affected by enzyme binding, the comparison of isotope fractionation would make more sense between the biological dehydrochlorination of the (+)-enantiomer (ε_c (+) of -2.4 ± 0.8 ‰ and -2.5 ± 0.6 ‰, Table 1) and chemical dehydrochlorination by alkaline hydrolysis (ε_c of -7.6 ± 0.4 ‰) (see discussion 4.3, mechanistic interpretation).

Similarly, dichloroelimination catalyzed by Fe^0 nanoparticles via two-electron transfers gave a slightly higher carbon isotope fractionation (ε_c of -4.9 ± 0.1 ‰) than dichloroelimination by anaerobic biodegradation (ε_c of -3.7 ± 0.8 ‰) (Badea et al., 2011). In both cases, α-HCH was transformed to TeCCH assuming that two C-Cl bonds

Table 1:. Comparison between AKIE values and typical (A)KIE values.

Reactions	Initial step	ε_c (‰) ± 95% CI		n	x	z	AKIE$_c$	(A)KIE$_c$
		$\varepsilon_c(+)$ (‰)	$\varepsilon_c(-)$ (‰)					
UV/H$_2$O$_2$ (≥ 280 nm)	Hydrogen abstraction (Nitoi et al., 2013)	-1.9 ± 0.2		6	1	1	1.012 ± 0.001	1.01 - 1.02(Elsner et al., 2005; Huskey, 1991)
		-1.7 ± 0.2	-2.1 ± 0.3					
Hydrolysis (pH 9.78)	Dehydrochlorination (Ngabe et al., 1993)	-7.6 ± 0.4		6	2	2	1.048 ± 0.003	1.03 - 1.09(Shiner and Wilgis, 1992)
		-7.2 ± 0.5	-7.7 ± 0.6					
UV (≥ 185 nm)	Dechlorination (McNaught and Wilkinson, 1997) Homolytic bond cleavage	-2.8 ± 0.2		6	1	1	1.017 ± 0.001	1.02 - 1.03 (Hofstetter et al., 2007)
		-2.4 ± 0.3	-2.7 ± 0.3					
Electrochemical reduction	Dechlorination (Houmam, 2008) Single-electron transfer	-3.8 ± 0.4		6	1	1	1.023 ± 0.003	1.02 - 1.03 (Hofstetter et al., 2007)
		-3.0 ± 0.3	-3.6 ± 0.2					
Reduction by Fe^0	Dichloroelimination (Elliott et al., 2009) Two-electron transfers	-4.9 ± 0.1		stepwise			1.030 ± 0.0006	1.01 - 1.03 (Fletcher et al., 2009)
				6	1	1		
		-5.1 ± 0.4	-4.8 ± 0.5	concerted			1.015 ± 0.0003	1.007 - 1.017 (Fletcher et al., 2009)
				6	2	1		
Anaerobic (C. pasterianum)	Dichloroelimination (Badea et al., 2011) Two-electron transfers	-3.7 ± 0.8$^\$$		stepwise			1.023 ± 0.005$^\$$	1.01 - 1.03 (Fletcher et al., 2009)
		-	-	6	1	1		
Aerobic (Sphingobium indicum strain B90A)	Dehydrochlorination (Suar et al., 2005a)	-1.6 ± 0.3*		6	2	2	AKIE$_c$(+) 1.015 ± 0.005 AKIE$_c$(-) 1.006 ± 0.004	1.0168 and 1.0218$^\$$ (Manna and Dybala-Defratyka, 2013)
		-2.4 ± 0.8*	-1.0 ± 0.6*					
Aerobic (Sphingobium japonicum strain UT26)	Dehydrochlorination (Bashir et al., 2013)	-1.0 ± 0.2*		6	2	2	1.015 ± 0.004 1.004 ± 0.001	1.0168 and 1.0218$^\$$ (Manna and Dybala-Defratyka, 2013)
		-2.5 ± 0.6*	-0.7 ± 0.2*					

$^\$$Badea et al. (2011) (Badea et al., 2011); *Bashir et al. (2013) (Bashir et al., 2013); $^\$$Data from quantum mechanical modelling (Manna and Dybala-Defratyka, 2013).

were cleaved in a stepwise or a concerted mechanism. However, the biological process yielded a lower ε_c which was most probably caused by non-fractionating processes prior to the isotope sensitive bond cleavage (e.g. binding of the substrate to the enzyme). Since the ε_c of α-HCH for chemical transformation was systematically higher compared to biological transformation, the extent of carbon isotope fractionation may allow distinguishing biological and chemical reactions.

Apparent kinetic isotope effect (AKIE)

Carbon isotope enrichment factors (ε_c) were converted to AKIEC for the different reaction mechanisms of α-HCH. AKIEC values were calculated as described in SI and compared with putative kinetic carbon isotope effects (KIEC). While the number of carbon atoms of α-HCH (n) was the same for all reactions, different values for the number of reactive positions (x) and the number of positions in intra-molecular competition (z) were taken into account depending on the reaction scenario (Table 1). Semiclassical Streitwieser Limits for KIEC of C–H bond cleavage were derived in the range of 1.01 to 1.02 (Elsner et al., 2005; Huskey, 1991). The AKIEC value was calculated as 1.012 ± 0.001 for the OH radical-induced H abstraction during the UV/H2O2 process, corresponding well to the expected range.

Calculated AKIEC of 1.048 ± 0.003 for dehydrochlorination via alkaline hydrolysis was consistent with the (A)KIEC range of 1.03 to 1.09 expected for simultaneous cleavage of a C–H and a C–Cl bond (Shiner and Wilgis, 1992) by a bimolecular elimination (E2 reaction). For biological dehydrochlorination, AKIEC values for (+)-enantiomer (see discussion 4.2) were calculated as 1.015 ± 0.005 and 1.015 ± 0.004, respectively. These values were significantly smaller than AKIEC for alkaline hydrolysis indicating that the mode of bond cleavage via E2 reaction might have different transition states between the chemical and biological dehydrochlorination. Quantum mechanical modelling of α-HCH biotransformation by LinA resulted in KIEC of 1.0168 and 1.0218 at reactive positions (Manna and Dybala-Defratyka, 2013). These values corresponded well to AKIEC values obtained from experimental biotransformation, suggesting that a complex transition state was governing the biological dehydrochlorination.

AKIE$_C$ calculated for direct photolysis via homolytic cleavage of a C–Cl bond (1.017 ± 0.001) was in good agreement with KIE$_C$ values for chemical dechlorination (1.02 to 1.03) (Hofstetter et al., 2007). Likewise, AKIE$_C$ calculated for electro-reduction (1.023

± 0.003) by single-electron transfer with a C-Cl bond cleavage was in the same expected range of 1.02 to 1.03 for previously reported chemical dechlorination (Hofstetter et al., 2007) and, thus, supported a reductive dechlorination pathway.

For Fe^0 nanoparticles-induced dichloroelimination, two C-Cl bonds were cleaved via a stepwise mode or a concerted mode. Therefore, $AKIE_C$ values for dichloroelimination were determined as 1.030 ± 0.0006 and 1.015 ± 0.0003, assuming a stepwise or a concerted pathway, respectively. These values were within the respective range of 1.01 to 1.03 and 1.007 to 1.017 for previously reported $AKIE_C$ of dichloroelimination reactions (Fletcher et al., 2009).

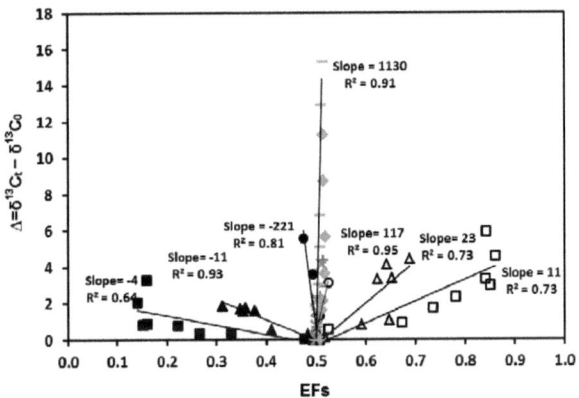

Figure 3: Comparison of carbon isotope discrimination ($\delta^{13}C_t - \delta^{13}C_0$) vs enantiomeric fraction (EF) for (+) α-HCH (EF (+)) (open symbols) and (−) α-HCH (EF (−)) (closed symbols) for anaerobic biodegradation by Clostridium pasterianum (circles) and aerobic biodegradation by Sphingobium indicum strain B90A (squares) and Sphingobium japonicum strain UT26 (triangles). Colored symbols stand for chemical reactions: direct photolysis (+), indirect photolysis (−), hydrolysis (−), electrochemical reduction (×) and reduction by Fe^0 (♦). The slope is only shown for hydrolysis due to the overlap with other chemical reactions.

Enantiomeric fraction and isotope discrimination

Recently, enantiomeric fraction and enantioselective carbon isotope fractionation have been investigated for aerobic/anaerobic biotransformation of the α-HCH enantiomers using ESIA (Bashir et al., 2013). Enantiomeric fraction (EF) (+) and EF (−) described in SI were plotted against the carbon isotope discrimination ($\Delta = \delta_t - \delta_0$) of the (+)-enantiomer and (−)-enantiomer, respectively (Figure 3). It is shown that the ESIA approach is suitable to differentiate between chemical transformation and aerobic biodegradation. However, the range of chemical transformation partially overlaps with those of anaerobic biodegradation and, thus, the distinction of these transformation processes seems to be not always possible at least when low extent of transformation does not allow to quantify the isotope fractionation due to the analytical uncertainty.

Conclusion

The specific carbon isotope fractionations for the different reaction mechanisms obtained in our study can, thereby, be used as references for evaluating field data in order to determine the relevance of chemical and biological α-HCH transformation in the environment. Moreover, our study reveals that CSIA in combination with ESIA and the determination of EF values is feasible as a generic concept for characterizing the transformation of chiral contaminants. However, limitations appear to distinguish reactions with low enantioselectivity for instance between chemical transformation and anaerobic biodegradation of α-HCH. In the future, it is envisioned that CSIA will be available for measuring chlorine ($^{37}Cl/^{35}Cl$) and hydrogen ($^{2}H/^{1}H$) isotope signatures of HCH. Hence, it should be possible to apply the multi-dimensional CSIA for a better characterization of transformation pathways, e.g. between C–H and C–Cl bond cleavage involved reactions. This will further open up new perspectives for detailed evaluation of the fate of α-HCH.

Acknowledgments

We thank Ursula Günther, Matthias Gehre and Falk Bratfisch for technical support of isotope analysis. This research has been financially supported by the European

Union under the 7th Framework Programme (project acronym CSI: ENVIRONMENT, contract number PITN-GA-2010-264329), and supported by University of Agriculture Faisalabad, Pakistan and the Helmholtz Impulse and Networking Fund through Helmholtz Interdisciplinary Graduate School for Environmental Research (HIGRADE).

References

Abhilash P., Singh N. (2009) Pesticide use and application: An Indian scenario. Journal of Hazardous Materials, 165(1):1-12.

Aelion C. M., Höhener, P., Hunkeler, D., & Aravena, R. (Eds.). (2009).Environmental isotopes in biodegradation and bioremediation, 1st edn. CRC Press, Boca Raton.

Badea S.L., Vogt C., Weber S., Danet A.-F., Richnow H.-H. (2009) Stable isotope fractionation of γ-hexachlorocyclohexane (Lindane) during reductive dechlorination by two strains of sulfate reducing bacteria. Environmental Science & Technology, 43(9):3155-3161.

Badea S.L., Vogt C., Gehre M., Fischer A., Danet A.F., Richnow H.H. (2011) Development of an enantiomer- specific stable carbon isotope analysis (ESIA) method for assessing the fate of α-hexachlorocyclohexane in the environment. Rapid Communications in Mass Spectrometry, 25(10):1363-1372.

Bashir S., Fischer A., Nijenhuis I., Richnow H.-H. (2013) Enantioselective carbon stable isotope fractionation of hexachlorocyclohexane during aerobic biodegradation by *Sphingobium* spp. Environmental Science & Technology, 47(20):11432-11439.

Bhatt P., Kumar M.S., Chakrabarti T. (2007) Assessment of bioremediation possibilities of technical grade hexachlorocyclohexane (tech-HCH) contaminated soils. Journal of Hazardous Materials, 143(1):349-353.

Bidleman T.F., Jantunen L.M., Kurt-Karakus P.B., Wong F. (2012) Chiral persistent organic pollutants as tracers of atmospheric sources and fate: review and prospects for investigating climate change influences. Atmospheric Pollution Research, 3(4):371-382.

Elliott D.W., Lien H.-L., Zhang W.-X. (2008) Zerovalent iron nanoparticles for treatment of ground water contaminated by hexachlorocyclohexanes. Journal of Environmental Quality, 37(6):2192-2201.

Elliott D.W., Lien H.-L., Zhang W.-X. (2009) Degradation of lindane by zero-valent iron nanoparticles. Journal of Environmental Engineering, 135(5):317-324.

Elsner M., Zwank L., Hunkeler D., Schwarzenbach R.P. (2005) A new concept linking observable stable isotope fractionation to transformation pathways of organic pollutants. Environmental Science & Technology, 39(18):6896-6916.

Elsner M., Cwiertny D.M., Roberts A.L., Lollar B.S. (2007) 1,1,2,2-tetrachloroethane reactions with OH-, Cr(II), granular iron, and a copper-iron bimetal: Insights from product formation and associated carbon isotope fractionation. Environmental Science & Technology, 41(11):4111-4117.

Evans D.H. (2008) One-electron and two-electron transfers in electrochemistry and homogeneous solution reactions. Chemical Reviews, 108(7):2113-2144.

Fiedler H., Hub M., Hutzinger O. (1993) Stoffbericht hexachlorcyclohexan (HCH) LfU.

Fletcher K.E., Löffler F.E., Richnow H.-H., Nijenhuis I. (2009) Stable carbon isotope fractionation of 1, 2-dichloropropane during dichloroelimination by *Dehalococcoides* populations. Environmental Science & Technology, 43(18):6915-6919.

Fu F., Dionysiou D.D., Liu H. (2014) The use of zero-valent iron for groundwater remediation and wastewater treatment: A review. Journal of Hazardous Materials, 267:194-205.

Hamada M., Kawano E., Kawamura S., Shiro M. (1982) A new isomer of 1,2,3,4,5-pentachlorocyclohexane from UV irradiation products of alpha-isomers, beta-isomers and delta-isomers of 1,2,3,4,5,6-hexachlorocyclohexane. Agricultural and Biological Chemistry, 46(1):153-157.

Hofstetter T.B., Reddy C.M., Heraty L.J., Berg M., Sturchio N.C. (2007) Carbon and chlorine isotope effects during abiotic reductive dechlorination of polychlorinated ethanes. Environmental Science & Technology, 41(13):4662-4668.

Houmam A. (2008) Electron transfer initiated reactions: Bond formation and bond dissociation. Chemical Reviews 108:2180-2237.

Hühnerfuss H., Faller J., Kallenborn R., König W.A., Ludwig P., Pfaffenberger B., Oehme M., Rimkus G. (1993) Enantioselective and nonenantioselective degradation of organic pollutants in the marine ecosystem. Chirality, 5(5):393-399.

Huskey W.P. (1991) Origins and interpretations of heavy-atom isotope effects, in: P. F. Cook (Ed.), Enzyme mechanism from isotope effects, CRC Press: Boca Raton, FL. pp. 37-72.

Jiang Y.F., Wang X.-T., Jia Y., Wang F., Wu M.-H., Sheng G.-Y., Fu J.-M. (2009) Occurrence, distribution and possible sources of organochlorine pesticides in agricultural soil of Shanghai, China. Journal of Hazardous Materials, 170(2):989-997.

Lal R., Pandey G., Sharma P., Kumari K., Malhotra S., Pandey R., Raina V., Kohler H.-P.E., Holliger C., Jackson C. (2010) Biochemistry of microbial degradation of hexachlorocyclohexane and prospects for bioremediation. Microbiology and Molecular Biology Reviews, 74(1):58-80.

Lallas P.L. (2001) The Stockholm Convention on persistent organic pollutants. American Journal of International Law, 95:692-708.

Li Y.F. (1999) Global technical hexachlorocyclohexane usage and its contamination consequences in the environment: from 1948 to 1997. Science of the Total Environment, 232(3):121-158.

Liu X.M., Peng P.A., Fu J.M., Huang W.L. (2003) Effects of FeS on the transformation kinetics of gamma-hexachlorocyclohexane. Environmental Science & Technology, 37(9):1822-1828.

Manna, R. N., & Dybala-Defratyka, A. (2013). Insights into the elimination mechanisms employed for the degradation of different hexachlorocyclohexane isomers using kinetic isotope effects and docking studies. Journal of Physical Organic Chemistry, 26(10), 797-804.

McNaught A.D., Wilkinson A. (1997) Compendium of chemical terminology Blackwell Science Oxford.

Murthy, H. M., & Manonmani, H. K. (2007). Aerobic degradation of technical hexachlorocyclohexane by a defined microbial consortium. Journal of Hazardous Materials, 149(1), 18-25.

Ngabe B., Bidleman T.F., Falconer R.L. (1993b) Base hydrolysis of α- and γ-Hexachlorocyclohexanes. Environmental Science & Technology, 27(9):1930-1933.

Nienow A.M., Bezares-Cruz J.C., Poyer I.C., Hua I., Jafvert C.T. (2008) Hydrogen peroxide-assisted UV photodegradation of Lindane. Chemosphere, 72(11):1700-1705.

Nijenhuis I., Andert J., Beck K., Kästner M., Diekert G., Richnow H.-H. (2005) Stable isotope fractionation of tetrachloroethene during reductive dechlorination by *Sulfurospirillum multivorans* and *Desulfitobacterium* sp. strain PCE-S and abiotic reactions with cyanocobalamin. Applied and Environmental Microbiology, 71(1):3413-3419.

Nitoi I., Oncescu T., Oancea P. (2013) Mechanism and kinetic study for the degradation of lindane by photo-fenton process. Journal of Industrial and Engineering Chemistry, 19(1):305-309.

Northrop D.B. (1981) The expression of isotope effects on enzyme-catalyzed reactions. Annual Review of Biochemistry, 50(1):103-131.

Peng P., Ren M.Z., Huang W.L. (2004) Carbon isotope fractionation during the hydrolytic reactions of α- and γ-hexachlorocyclohexanes (HCHs). In Abstracts of Papers of the American Chemical Society, 228:U605-U605.

Ruiz M.S., Yu A.S., Martins F.E. (2012) Remediation technologies for areas contaminated with organochlorine: A preliminary assessment of their worldwide application based on a literature review, Technology Management for Emerging Technologies (PICMET), 2012 Proceedings of PICMET'12:, IEEE. pp. 308-318.

Shiner V.J., Wilgis F.P. (1992) Heavy atom isotope rate effects in solvolytic nucleophillic reactions at saturated carbon, in: E. Buncel and W. H. Saunders (Eds.), Isotopes in organic chemistry, Amsterdam: Elsevier. pp. 239-335.

Singh R., Misra V., Singh R.P. (2011) Remediation of-hexachlorocyclohexane contaminated soil using nanoscale zero-valent iron. Journal of Bionanoscience, 5(1):82-87.

Singh R., Manickam N., Mudiam M.K.R., Murthy R.C., Misra V. (2013) An Integrated (Nano-Bio) Technique for Degradation of γ-HCH Contaminated Soil. Journal of Hazardous Materials, 258:35-41.

Srivastava A.P.M.R. (2006) Electrochemical reduction of Lindane (-HCH) at $NiCo_2O_4$ modified electrode. Microbiology, 102:1468-1478.

Suar M., Dogra C., Raina V., Kohler H., Poiger T., Hauser A., Buser H., van der Meer J., Holliger C., Lal R. (2005a) Enantioselective transformation of α-HCH by dehydrochlorinase (LinA1, LinA2) from *Sphingomonas paucimobilis* B90A. Applied and Environmental Microbiology, 71(12):8514-8518.

Suar M., Hauser A., Poiger T., Buser H.R., Muller M.D., Dogra C., Raina V., Holliger C., van der Meer J.R., Lal R., Kohler H.P.E. (2005b) Enantioselective transformation of α-hexachlorocyclohexane by the dehydrochlorinases LinA1 and LinA2 from the soil bacterium Sphingomonas paucimobilis B90A. Applied and Environmental Microbiology, 71:8514-8518.

Tobiszewski M., Namiesnik J. (2012) Abiotic degradation of chlorinated ethanes and ethenes in water. Environmental Science and Pollution Research, 19:1994-2006.

Vijgen J. (2006) The legacy of lindane HCH isomer production. Main Report, IHPA, Holte, January.

Von Sonntag C., Schuchmann H.P. (1991) The elucidation of peroxyl radical reactions in aqueous solution with the help of radiation-chemical methods. Angewandte Chemie International Edition in English 30(10):1229-1253.

Wang C.B., Zhang W.X. (1997) Synthesizing nanoscale iron particles for rapid and complete dechlorination of TCE and PCBs. Environmental Science & Technology, 31(7):2154-2156.

Wang Z., Peng P.a., Huang W. (2009) Dechlorination of γ-hexachlorocyclohexane by zero-valent metallic iron. Journal of Hazardous Materials, 166(2):992-997.

Zaleska A., Hupka J., Wiergowski M., Biziuk M. (2000) Photocatalytic degradation of lindane, p,p'-DDT and methoxychlor in an aqueous environment. Journal of Photochemistry and Photobiology A-Chemistry, 135(2-3):213-220.

Zhang W.-X. (2003) Nanoscale iron particles for environmental remediation: an overview. Journal of Nanoparticle Research, 5(3-4):323-332.

Zhang W.X., Wang C.B., Lien H.L. (1998) Treatment of chlorinated organic contaminants with nanoscale bimetallic particles. Catalysis Today, 40(4):387-395.

3.4

Evaluating Degradation of Hexachlorocyclohexane (HCH) Isomers within a Contaminated Aquifer using Compound Specific Stable Carbon Isotope Analysis

Safdar Bashir[t], Kristina L. Hitzfeld[t], Matthias Gehre[t], Hans-Hermann. Richnow[t], Anko Fischer[t§]

[t]Helmholtz Centre for Environmental Research - UFZ, Department of Isotope Biogeochemistry, Permoserstr. 15, D-04318 Leipzig, Germany

[§]Isodetect GmbH - Company for isotope monitoring, Deutscher Platz 5b, D-04103 Leipzig, Germany

Chapter : 3.4 Results

Abstract

The applicability of carbon stable carbon isotope analysis (CSIA) for source identification and assessment of biodegradation of hexachlorocyclohexane (HCH) isomers was investigated in a contaminated aquifer at a former pesticide processing facility. A CSIA method was developed and tested for efficacy in determining carbon isotope ratios of HCH isomers in groundwater samples using gas chromatography coupled with isotope ratio mass spectrometry (GC-IRMS). The method was able to confirm known HCH contaminant source zones near former processing facilities and a waste dumping site, as well as detect enrichment in carbon isotope ratios downstream of the contaminant sources, indicating that biodegradation of HCHs was underway. CSIA from monitoring campaigns in 2008 and 2010 revealed temporal trends in HCH biodegradation, providing information on the remediation efficacy of the natural attenuation process within specific zones of the investigated aquifer. Conservative calculations based on the Rayleigh equation revealed levels of HCH biodegradation ranging from 16-86 %. Moreover, time- and distance-dependent *in situ* first-order biodegradation rate constants were estimated, with maximal values of 0.0029 d^{-1} and 0.0098 m^{-1} for α-HCH, 0.0110 d^{-1} and 0.0370 m^{-1} for β-HCH, and 0.0058 d^{-1} and 0.0192 m^{-1} for δ-HCH, respectively. This study highlights the applicability of CSIA for source identification and assessment of HCH degradation within contaminated aquifers.

3.4.1 Introduction

Hexachlorocyclohexane (HCH) as Contaminants of Concern

Subsurface water contamination is an issue of growing concern because of the hazard it poses to important drinking water resources and subsurface water bodies (EEA, 2007; Ritter et al., 2002; Schwarzenbach et al., 2010; Zoumis et al., 2001). Persistent organic pollutants (POPs) form a major contaminant group in subsurface compartments and pose substantial environmental and health risks (Minh et al., 2006; Weber et al., 2011; Weber et al., 2008) Hexachlorocyclohexane (HCH) are globally dispersed POPs associated with the production and application of pesticides and are responsible for substantial environmental impacts via widespread contamination of soil and groundwater (Bhatt et al., 2009; Breivik et al., 1999; Heinisch et al., 2005; Li, 1999; Vijgen et al., 2011; Walker et al., 1999). Due to their combination of toxicity and environmental persistence, the commercial production and use of HCHs has been regulated by the Stockholm Convention since 2004 and the production of the three main HCH isomers (α, β & γ-HCH) has been banned since 2009 (Alvarez et al., 2012). However, while these measures serve to limit future risk from HCH contamination, significant risk still exists today due to the extensive historical use of HCHs, as well as the ongoing production and application of Lindane (γ-HCH) in several countries, remaining stockpiles from previous manufacturing, and leachates from earlier stockpiles into groundwater; all of which necessitate remediation strategies for HCH-contaminated sites (Bhatt et al., 2009; Vijgen et al., 2011).

Monitored Natural Attenuation (MNA)

One such strategy is Monitored Natural Attenuation (MNA), sometimes known as intrinsic remediation. MNA strategies consist of allowing natural degradation processes to occur while regularly monitoring contaminant levels to assess both the efficacy of remediation and current contaminant-derived risk (Bombach et al., 2010; Schirmer et al., 2006). In order to manage environmental risks effectively, MNA strategies require that appropriate monitoring tools be developed to assess *in situ* biodegradation (Bombach et al., 2010; Illman and Alvarez, 2009; Schirmer et al., 2006; Soga et al., 2004; Wiedemeier et al., 1999). Biodegradation is a key sustainable removal process of HCHs in soil- and aquifer systems (Bhatt et al., 2009), and a cost efficent alternative to physico-chemical treatment for HCH removal from groundwater and soils (Alvarez et al., 2012; Bombach et al., 2010; Langenhoff et al., 2013; Phillips et al., 2006). The microbial transformation of HCH has been extensively reviewed and variable degradation rates have been reported for all HCH isomers under various environmental conditions (Lal et al., 2010; Mehboob et al., 2013). The molecular structure of the HCH isomers plays a key role in biotransformation as it has been shown that, α and γ isomers are degraded faster than β and δ isomers (Langenhoff et al., 2013; Li et al., 2011; Phillips et al., 2006; Quintero et al., 2005). However, concentration-based assessment of biodegradation under *in situ* conditions is complicated due to various factors such as the transport, volatilization, dilution, and dispersion of contaminants (Bombach et al., 2010; Hatzinger et al., 2013).

In Situ Monitoring using CSIA

One of the most promising tools for both characterizing contaminant sources and monitoring *in situ* degradation of organic contaminants in aquifers is compound-specific stable isotope analysis, in which the stable isotope ratios of one or more elements in a given compound are measured in order to investigate the transformation processes at work. (Aelion et al., 2010; Elsner, 2010; Hofstetter and Berg, 2011; Meckenstock et al., 2004; Thullner et al., 2012). The reactions which together constitute degradation processes often result in changes in stable carbon isotope ratios ($^{13}C/^{12}C$) of pollutants (Meckenstock et al., 2004). Molecules with light carbon isotopes (^{12}C) in the reactive position require less energy for bond cleavage and thus tend to be degraded more readily than molecules containing heavy carbon isotopes (^{13}C), resulting in a differential process whereby degradation products become comparatively enriched in light isotopes and yet-to-be-degraded material becomes comparatively enriched in heavy isotopes, as the lighter isotopes are preferentially selected for degradation. This process is called stable isotope fractionation and can be detected via an enriched ratio of heavy isotopes (^{13}C) in the remaining stock of the pollutant compound. Biodegradation is associated with fractionation, therefore CSIA allows for the characterization of biodegradation activity based on the degree of fractionation found to have occurred, computed from the difference between the measured heavy-isotope enrichment of the remaining pollutant and the standard background isotope ratios of the element. In laboratory studies, statistically significant carbon isotope fractionation has been shown for HCH biodegradation under both oxic and anoxic conditions, and has been found to be more pronounced for anaerobic than for aerobic biodegradation (Badea et al., 2009; Badea et al., 2011; Bashir et al., 2013). However, the applicability of CSIA has not previously been demonstrated for *in situ* evaluation of sources or for the measurement of biodegradation at a HCH-contaminated field site. In this study, CSIA was therefore applied to investigate the origin and monitor the fate of HCHs within a contaminated aquifer in order to appraise the applicability of CSIA for assessing the remediation of HCHs in groundwater systems. The degradation of HCHs was investigated using CSIA during two monitoring campaigns taking place in 2008 and 2010 in order to determine the progress and sustainability of HCH degradation in the investigated aquifer. To our best knowledge, this is the first study which addresses the use of CSIA for assessing the fate of HCHs within a groundwater system.

3.4.2 Materials and Methods
Field site
Site history
The field site is located in and around a former pesticide formulating plant that included both a formulation site and a packaging facility. Pesticide formulation at this site began in 1935 and continued for more than five decades. The pesticides prepared at the site were mainly HCH-based but also included others such as dichlorodiphenyltrichloroethane (DDT). In addition to pesticides, solvents and chemicals for organic synbook (e.g., benzene, chlorinated aliphatic and aromatic compounds) were also stored and used at this facility. As known from historical information, HCH was not itself produced on-site, but technical HCH was instead purchased from suppliers for use in pesticide formulation. Losses of HCH-containing raw materials and products during purification, pesticide formulation, and storage, as well as irrigation and dumping of production-related wastes have led to extensive HCH contamination of soil and groundwater in several areas of the field site (Figure 1). At the field site, most relevant HCH isomers are α-, β-, γ-, and δ-HCH.

Geology and groundwater table
The highest pollutant concentrations were found within the upper quaternary aquifer, which consists of 12 to 15 m thick (glacio-) fluvial sand and gravel deposits. This aquifer is largely separated from the underlying tertiary aquifer by a 30 m thick clay- and coal-bed layer. The lower aquifer exhibits almost no contamination. The mean effective groundwater flow velocity of the upper aquifer was estimated at 0.3 m/d. The matrix of the upper aquifer displayed a very low organic carbon content of only 0.014 %, thus negligible retardation of HCHs was expected (Lotse et al., 1968).

In the early decades of the 20th century, drainage measures were initiated to facilitate nearby mining activity, which led to a lowering of the groundwater table and a reversal of groundwater flow direction in the area of the field site. Groundwater resurgence and realignment to the previous groundwater flow direction have occurred since the termination of these measures beginning in the early 1990s. During the last decade, the groundwater table has been relatively stable with a slight increasing tendency. The main direction of groundwater flow was largely constant toward north/northeast to north/northwest (Figure 1). Hence, the contaminant plume in the upper aquifer was established in the main groundwater flow direction and has achieved quasi-stationary conditions (Figure 1).

Monitoring wells

In order to assess sources and detect degradation of HCHs, carbon isotope ratios and concentrations of HCHs as well as hydro-geochemical parameters were determined for a transect of six wells located within the main groundwater flow direction (wells A-F, Figure 1) and sampled in 2008 and 2010.

Sampling

Groundwater sampling was carried out by an authorized contractor based on standard procedures given in the Supporting Information (SI) (Appendix A4). During groundwater monitoring campaigns in autumn 2008 and 2010, samples were collected from wells with filter screens spanning the entire width of the water-saturated zone within the upper aquifer using a submersible electrical pump. Samples for concentration and hydro-geochemical analysis were then sent to an analytical laboratory where they were processed immediately. For carbon isotope analysis of HCHs, two 1 L glass bottles (Schott, Germany) were filled with sampled groundwater and sealed with Teflon-coated caps (Schott, Germany) without headspace, in order to avoid evaporation. The groundwater samples were then adjusted to a pH of 2 using hydrochloric acid (HCl; 25 %, Carl Roth GmbH & Co. KG, Germany) to inhibit microbial activity. Groundwater samples for isotope analysis were stored in a dark environment at 4°C until extraction could take place.

Analytical procedures

Concentration analysis

Concentrations of dissolved oxygen, temperature, pH, redox potential, and electrical conductivity were measured during sampling using appropriate electrodes (CellOx® 325, SenTix® 41, SenTix® ORP, KLE 325; WTW GmbH, Germany).

Concentration analyses of contaminants and hydro-geochemical parameters were performed according to analytical standard procedures summarized in SI.

2.3.2 Carbon stable isotope measurements

For the carbon stable isotope analysis of HCHs, three extractions were taken from the 1L groundwater samples from each monitoring well, with 30 mL dichloromethane (DCM; ≥ 99.8 %, Carl Roth GmbH & Co. KG, Germany) in a separating funnel. All DCM extracts were combined and dried with anhydrous sodium sulfate (Na_2SO_4; ≥ 99 %, Bernd Kraft GmbH, Germany). The combined DCM extracts were reduced to approximately 1 mL using a rotary evaporator. The extraction procedure did not itself result in significant changes in carbon isotope ratios of HCHs, as verified and described in SI.

The carbon isotope ratios of HCHs were measured by gas chromatography - isotope ratio mass spectrometry (GC-IRMS), using a system described elsewhere (Badea et al., 2009; Badea et al., 2011). Detailed information on the GC-IRMS analysis of HCHs can be found in the SI. Quality control was carried out using isotope laboratory standards consisting of pure HCH isomers (97-99 %, Sigma-Aldrich Chemie GmbH, Germany) with known carbon isotope ratios determined by elemental analyzer - isotope ratio mass spectrometry (EA-IRMS) using reference materials (IAEA-CH-6, IAEA-CH-7) from the International Atomic Energy Agency (IAEA) (Coplen et al., 2006). The carbon isotope ratios were expressed in delta notation ($\delta^{13}C$) relative to the international standard Vienna Pee Dee Belemnite (V-PDB) according to eq. 1 (Coplen, 2011).

$$\delta^{13}C_{sample} = \frac{R_{sample}}{R_{standard}} - 1 \qquad (1)$$

R_{sample} and $R_{standard}$ are the $^{13}C/^{12}C$ ratios of the sample and V-PDB standard, respectively. $\delta^{13}C$-values were reported in percent per mil (‰). All samples were measured in at least triplicate. The total analytical uncertainty, incorporating both accuracy and reproducibility, was less than ±1.0 ‰ in almost all cases, as described in SI.

Quantitative interpretation of isotope data

The Rayleigh equation can be applied to mathematically describe microbial isotope fractionation processes, as shown in eq. 2:

$$\frac{(\delta_t + 1)}{(\delta_0 + 1)} = \left(\frac{C_{Bt}}{C_0}\right)^{\varepsilon} \qquad (2)$$

where δ_t is the isotope ratio of the substrate at a certain time t of biodegradation, δ_0 is the initial isotope ratio of the substrate, C_{Bt}/C_0 is the fraction of substrate remaining during biodegradation at a certain time t, and ε is the isotope enrichment factor (Mariotti et al., 1981).

The degree of contaminant biodegradation can be expressed as the percentage of the initial contaminant concentration decreased due to biodegradation (B [%]), as shown in eq. 3:

$$B[\%] = \left(1 - \frac{C_{Bt}}{C_0}\right) \cdot 100 \qquad (3)$$

Combining equation 3 with the Rayleigh equation (eq. 2) allows for the quantification of contaminant biodegradation over a time or distance interval (such as a groundwater flow path) using measured isotope ratios (Thullner et al., 2012). Required data inputs are the initial isotope ratio of the contaminant at a starting point in either time or in space (generally either the geographical contaminant source or the initial isotope ratio, depending on the frame of the analysis), and the isotope ratio of the remaining contaminant at a temporal or spatial observation point (e.g., a well downstream of the source). The amount of contaminant degraded between the starting point and observation point (x) is then given by equation 4 (Thullner et al., 2012):

$$B[\%] = \left(1 - \frac{C_{Bx}}{C_0}\right) \cdot 100 = \left[1 - \left(\frac{\delta_x + 1}{\delta_0 + 1}\right)^{\left(\frac{1}{\varepsilon}\right)}\right] \cdot 100 \qquad (4)$$

Moreover, *in situ* first-order biodegradation rate constants (λ_s) can be estimated using changes in isotope ratios over the distance between the initial and observation points using a Rayleigh-equation based approach (Hunkeler et al., 2008):

$$\lambda_s = -\frac{1}{\varepsilon \cdot s} \ln\left(\frac{\delta_x + 1}{\delta_0 + 1}\right) \qquad (5)$$

Time-dependent *in situ* first-order biodegradation rate constants (λ_t) can be determined by taking into account the travel time of the pollutants along the groundwater flow path:

$$\lambda_t = -\frac{1}{\varepsilon \cdot t} \ln\left(\frac{\delta_x + 1}{\delta_0 + 1}\right) \qquad (6)$$

where the travel time (t) can be approximated using the groundwater flow velocity (v) and the distance between the initial and observation points:

$$t \approx \frac{s}{v} \tag{7}$$

First-order biodegradation rate constants can be used to calculate biological half-life distances ($s_{1/2}$) or times ($t_{1/2}$), indicating the distance or time needed for the biodegradation of half of the initial pollutant concentration (Wiedemeier et al., 1999):

$$S_{1/2} = \frac{\ln 2}{\lambda_s} \tag{8}$$

$$t_{1/2} = \frac{\ln 2}{\lambda_t} \tag{9}$$

3.4.3 Results and Discussion
HCH distribution and hydro-geochemical conditions
Concentration of HCHs

Well A exhibited very high HCH concentrations, with values of 68 µg/L in 2008 and 48 µg/L in 2010, indicating a contaminant source in the immediate vicinity of these wells (Figure 1, Figure SI 3, supporting information, Appendix A4). Downstream of well A, decreasing HCH concentrations were observed at wells B and C on both sampling campaigns, with well B showing values of 38 µg/L in 2008 and 33 µg/L in 2010 and well C 58 µg/L and 10 µg/L in 2008 and 2010, respectively. (Figure 1, Figure SI 3, supporting information, Appendix A4). A significant increase compared to well C in HCH concentrations was detected at well D (110 µg/L in 2008, 91 µg/L in 2010) and at well E (196 µg/L in 2008, 72 µg/L in 2010) implying an additional source zone at these wells (Figure 1, Figure SI 3, supporting information, Appendix A4). Further downstream at well F, HCH concentrations displayed a decreasing trend, with values of 7 µg/L in 2008 and 3 µg/L in 2010 (Figure 1, Figure SI 3, supporting information, Appendix A4).

HCH Isomers

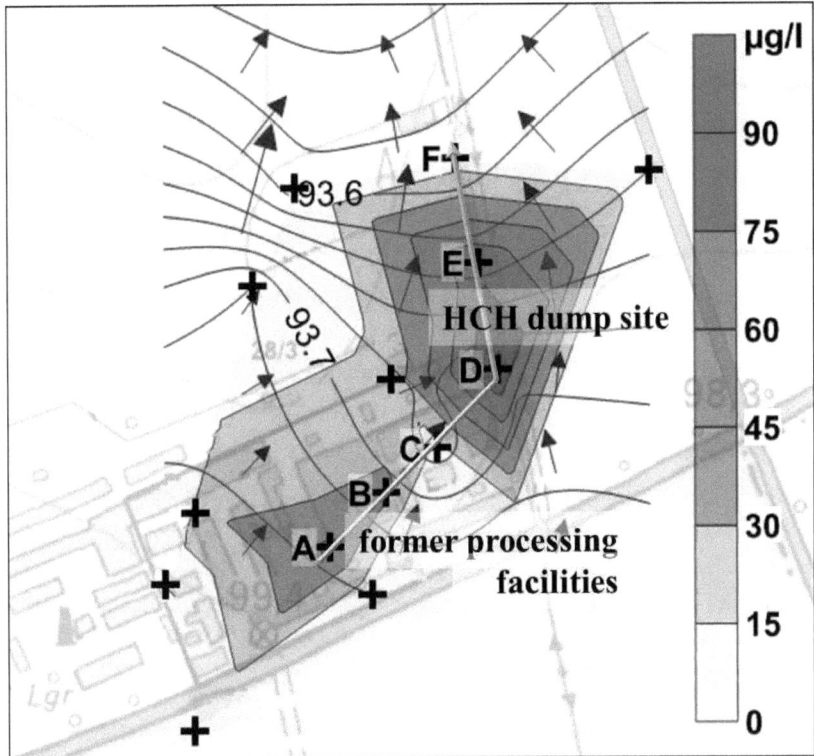

Figure 1: Distribution of HCHs (sum concentrations [µg/L] in 2010) and groundwater flow direction (blue arrows) within the upper aquifer of the investigated field site as well as historical site information. Crosses show locations of groundwater wells. The white large arrow indicates the investigated transect of wells A to F along the main groundwater flow direction.

At the field site, the most prevalent HCH isomers in the groundwater of the upper aquifer were α-, β-, γ-, δ-HCH. The highest concentration levels detected among all isomers were both for δ-HCH, with values of 120 µg/L at well E in 2008 and 67 µg/L at well D in 2010 (Figure 2), and at most wells, the detected concentrations of the other HCH isomers were at least three times lower than those of δ-HCH, with β-HCH generally more abundant compared to α- and γ-HCH (Figure 2). Thus, the concentration pattern of HCHs at the investigated field site revealed a familiar trend of biodegradability as generally observed in degradation experiments: δ-HCH=β-HCH > α-HCH > γ-HCH (Quintero et al., 2005).

Biogeochemical parameters

Reducing and anoxic conditions (redox potentials < -20 mV and oxygen ≤ 2 mg/L) were observed in the expected source zone at well A and in its downstream groundwater at wells B and C (Figure SI 4, supporting information, Appendix A4). The redox potential increased further downstream at wells D, E, and F, but oxygen concentrations remained at < 2 mg/L, implying that reducing and anoxic conditions prevail towards the fringe of the contaminant plume (Figure SI 4, supporting information, Appendix A4). At wells D and E, nitrate concentrations of > 30 mg/L detected in 2008 (Figure. SI 4, supporting information, Appendix A4) suggests that nitrate might be a relevant electron acceptor. High ammonium concentrations of 22 mg/L in 2008 and 14 mg/L in 2010 were measured at well D, perhaps caused by strong nitrate reduction or anthropogenic input (Figure. SI 4, supporting information, Appendix A4). Sulfate concentrations were higher in the source zone at well A, at 441 mg/L in 2008 and 382 mg/L in 2010, and decreased downstream at wells B and C, with values of 357 mg/L in 2008 and 309 mg/L in 2010 detected at well B and 358 mg/L in 2008 and 252 mg/L in 2010 at well C (Figure SI 4, supporting information, Appendix A4). This decrease could be attributable to sulfate reduction. At wells D, E, and F, sulfate concentrations were relatively high, at > 450 mg/L, thus sulfate seemed to be a negligible electron acceptor in the processes occurring in this area (Figure SI 4, supporting information, Appendix A4). High methane concentrations, measuring > 4000 µg/L, were also detected at well A (Figure. SI 4, supporting information, Appendix A4), suggesting the occurrence of methanogenesis within the contaminant source zone. Measured methane concentrations decreased with increasing distance from well A as well as with the passage of time from 2008 to 2010 (Figure SI 4, supporting information, Appendix A4). Relatively high methane concentrations (1800 µg/L in 2008, 1100 µg/L in 2010) were also detected for well F (Figure SI 4, supporting information, Appendix A4), suggesting distinct areas of methanogenesis at the fringe of the contaminant plume. Overall, HCH plume- and hydro-geochemical parameters indicated that reducing and anoxic conditions prevailed at the investigated transect along wells A to F.

Qualitative assessment of HCH degradation

The isotope data obtained from the monitoring campaigns in 2008 and 2010 was used to evaluate natural attenuation of HCHs within the investigated aquifer as well as temporal variations over the two-year period. Given the overall analytical error of ±1 ‰, differences in carbon isotope ratios needed to be at least ±2 ‰ to be considered significant. As δ-HCH showed the highest concentration values of all HCH isomers, it was considered particularly prevalent and was therefore the main focus for the interpretation of the obtained isotope data. Notable trends in isotope- and concentration patterns of the other HCH isomers are also subsequently examined.

δ-HCH

High δ-HCH concentrations of 57 µg/L in 2008 and 34 µg/L in 2010 were observed within the expected source zone of well A. At this well, δ-HCH exhibited carbon isotope ratios ($\delta^{13}C_{\delta\text{-HCH}}$) of -27.6 ‰ and -27.3 ‰ in 2008 and 2010, respectively (Figure 2). The similarity between these carbon isotope ratios indicates negligible fractionation, implying that the decrease in concentration over the two-year measurement period was caused primarily by physical processes rather than biodegradation. In the groundwater directly downstream of well A, samples obtained at well B displayed carbon isotope ratios of -27.0 ‰ in 2008 and -26.5 ‰ in 2010. This does reveal a tendency towards increasing $\delta^{13}C_{\delta\text{-HCH}}$-values, but the difference was not significant given the overall analytical error of ±1 ‰. Thus, the lower δ-HCH concentrations at well B (34 µg/L in 2008 and 29 µg/L 2010) compared to well A can also be attributed primarily to physical processes. Given that retardation of pollutants by sorption and evaporation can be neglected due to both the low organic matter content in the aquifer's matrix (measuring only 0.014 %) and the low tendency of volatilization of HCHs from water (Sahsuvar et al., 2003), dispersion and dilution are likely the most relevant processes for the decreased δ-HCH concentration. In comparison to well A, δ-HCH sampled at well C was more ^{13}C-enriched, with values of -26.3 ‰ in 2008 and -24.7 ‰ in 2010, however this ^{13}C-enrichment only passed the threshold of statistical significance in 2010 (Figure 2).

The ^{13}C-enrichment at well C was associated with decreasing δ-HCH concentrations, from 56 µg/L in 2008 to 8.6 µg/L in 2010, indicating a temporally-linked δ-HCH

Chapter : 3.4 Results

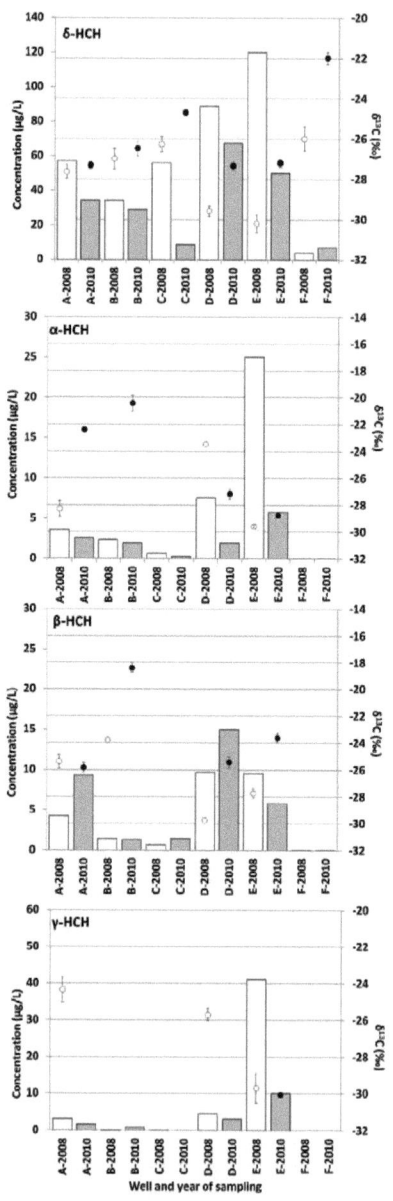

Figure 2: Concentrations and carbon isotope ratios of HCH isomers for the monitoring campaigns 2008 and 2010. Axes for concentrations and $\delta^{13}C$- values are partially different for the HCHs.

biodegradation at this well (Figure 2). However, in the groundwater even further downstream of well C, δ-HCH concentrations for both years increased significantly at wells D (89 µg/L in 2008 and 67 µg/L in 2010) and E (120 µg/L in 2008 and 50 µg/L in 2010). Quite surprisingly, in 2008, the $\delta^{13}C_{\delta\text{-HCH}}$-values at these wells were found to be even more ^{13}C-depleted than at well A, with -29.6 ‰ at well D and -30.2 ‰ at well E versus the aforementioned -27.6 ‰ at well A. By 2010, carbon isotope ratios of δ-HCH for wells D with -27.4 ‰ and E with -27.3 ‰ were found to have reached parity with well A. The higher δ-HCH concentrations and initially more negative $\delta^{13}C_{\delta\text{-HCH}}$-values for wells D and E compared to well A revealed a second discrete HCH source in the area of wells D and E (Figure 2). This finding was confirmed with historical maps, which revealed a former HCH dump in the vicinity of these wells, which explains both the spike in concentration and the presence of unfractionated-carbon bearing HCH (Figure 1).

The observed decrease in concentration from 2008 to 2010 at wells D and E was accompanied by ^{13}C-enrichment, yielding values of -27.4 ‰ at well D and -28.2 ‰ at well E. The change in $\delta^{13}C_{\delta\text{-HCH}}$-values was >+2 ‰ from 2008 to 2010 and indicates that biodegradation indeed contributed to the natural attenuation of δ-HCH in the source zone of wells D and E. At well F, located downstream of well D and E, δ-HCH showed lower concentrations, at 4 µg/L in 2008 and 7 µg/L in 2010, and was significantly more ^{13}C-enriched, at -26.0 ‰ in 2008 and -22.0 ‰ in 2010, providing evidence of δ-HCH biodegradation at the fringe of the contaminant plume (Figure 2).

Less-prevalent HCH isomers

Similar to δ-HCH, α-, β- and γ-HCH generally showed greater ^{13}C-enrichment as concentrations decreased, indicating sustained biodegradation of these HCHs (Figure 2, for more details see supporting information, Appendix A4). Moreover, these HCH isomers confirmed findings of HCH sources in the area of wells A and D/E. However, α-, β- and γ-HCH also displayed an inconsistent relationship between change in carbon isotope ratio and changes in concentration (see SI). In those cases, trends of change in concentrations of HCHs can provide information on natural attenuation or recharge of HCHs in groundwater but can provide only limited indications for biodegradation. By means of CSIA, biodegradation could be more precisely revealed.

Discussion of CSIA results

The carbon isotope ratios in conjunction with respective concentrations of HCHs revealed two main contaminant source zones located at wells A and D/E, respectively (Figure 2). The HCH source at well A resulted from contamination at former processing facilities and the source at wells D and E from previous dumping of HCH wastes (Figure 1). Carbon isotope analysis of HCHs provided evidence of HCH biodegradation in the groundwater at wells C and F, downstream of the main HCH source zones, indicating that biodegradation contributed to the natural attenuation of HCHs within the investigated aquifer (Figure1). From 2008 to 2010, concentrations of HCHs decreased concomitant with changes in carbon isotope ratios at most wells, indicating that the contribution of biodegradation for the natural attenuation of HCHs increased (Figure 2).

Quantification of HCH biodegradation

Using the Rayleigh-equation approach, pollutant degradation within groundwater systems can be estimated by changes in isotope ratios (Thullner et al., 2012). Calculations of the percentage of biodegradation (B [%]), distance-dependent and time-dependent *in situ* first-order biodegradation rate constants (λ_s [1/m], λ_t [1/d]) as well as half-life distances and times ($s_{1/2}$ [m], $t_{1/2}$ [d]) were carried out for expected flow paths within the main groundwater flow direction, showing changes in carbon isotope ratios of HCHs. In order to evaluate the sustainability of microbial removal, pollutant biodegradation was quantified based on isotope data obtained for monitoring campaigns in 2008 and 2010. In order to avoid overestimation of pollutant biodegradation, enrichment factors (ε_c) exhibiting the highest possible carbon isotope fractionation were used, thus providing the most conservative calculated estimates (Cichocka et al., 2008). Enrichment factors exhibiting maximum carbon isotope fractionation have previously been determined for the anaerobic biodegradation of HCHs (Badea et al., 2009; Badea et al., 2011; Bashir et al., 2013). Given that the contaminant plume exhibited mainly anoxic conditions (Figure SI 4, supporting information, Appendix A4), the choice of ε_c -values for anaerobic HCH biodegradation was reasonable for the calculations. Thus far, carbon isotope enrichment factors have only been determined for anaerobic α- and γ-HCH biodegradation with ε_c-values of -3.7 ± 0.8 ‰ (Badea et al., 2011) and -3.9 ‰ ± 0.6 (Badea et al., 2009; Badea et al., 2009), respectively. Similar reaction mechanisms can be expected for anaerobic biodegradation of all HCHs (Mehboob et al., 2013), which is confirmed by similarity of enrichment factors calculated by Badea et al. (2009, 2011). Thus, it can be safely assumed that all isomers will exhibit similar carbon isotope fractionation. Based on this assumption, the carbon isotope enrichment factor for anaerobic γ-HCH biodegradation exhibiting highest carbon isotope fractionation (ε_c = -3.9 ‰ (Badea et al., 2009)) was applied to calculate the biodegradation of α-, β- and δ-HCH.

Percentage of biodegradation

Due to low γ-HCH concentrations, only a limited number of $\delta^{13}C_{\gamma\text{-HCH}}$-values were obtained, thus it was not possible to quantify γ-HCH biodegradation. Biodegradation of α-HCH was calculated for the flow path from well A to well B (A→B) in 2010 and revealed biodegradation of 40 % (Table 1). For the same flow path, biodegradation percentages of 34 % in 2008 to 86 % in 2010 were calculated for β-HCH (Table 1), and revealed an increasing rate of β-HCH removal. Also for this flow path (A→B), δ-HCH exhibited low percentages of biodegradation, at only 16 % in 2008 and 19 % in 2010 (Table 1). Higher δ-HCH biodegradation was determined for the flow path from well A to well C in both years, at 30 % in 2008 and 50 % in 2010 (Table 1). Downstream of the second distinct source zone near wells D and E, the highest biodegradation percentage was obtained, with 67 % in 2008 and 75 % in 2010 along the path from well E to well F (E→F). Interestingly, biodegradation in 2010 was higher than in 2008 for all wells in the transect, indicating a general increase in the rate of HCH removal.

In situ first-order biodegradation rate constants

Time- and distance-dependent *in situ* first-order biodegradation rate constants (λ_t, λ_s) for α-HCH were estimated with 0.0029 d^{-1} and 0.0098 m^{-1} for the flow path from well A to well B (A→B) in 2010 (Table 1). For same flow path and year, higher biodegradation rate constants with 0.0111 d^{-1} and 0.0370 m^{-1} were obtained for β-HCH while its biodegradation rate constants in 2008 of 0.0024 d^{-1} and 0.0080 m^{-1} were similar to those for α-HCH in 2010. Compared to α- and β-HCH, biodegradation rate constants of δ-HCH were lower on the flow path from well A to well B (λ_t: 0.0010 - 0.0012 d^{-1}, λ_s: 0.0032 - 0.0042 m^{-1}) (Table 1). Thus, it can be inferred that δ-HCH was the most recalcitrant HCH isomer with respect to calculated biodegradation rate constants and half-life values (Table 1). A high recalcitrance of δ-HCH under anoxic conditions has been suggested in other studies (Jagnow et al., 1977; Quintero et al., 2005). Compared to flow path A→B, higher biodegradation rate constants for δ-HCH were estimated for flow paths A→C (λ_t: 0.0011 – 0.0021 d^{-1}, λ_s: 0.0037 – 0.0070 m^{-1}) and E→F (λ_t: 0.0047 – 0.0058 d^{-1}, λ_s: 0.0156 – 0.0192 m^{-1}), indicating more pronounced δ-HCH removal further downstream of the source zone at well A and on the fringe of the contaminant plume, respectively. Both flow paths exhibited an increase in biodegradation rate constants for 2010, which can be attributed to increasing δ-HCH removal. Overall, the λ_t-values obtained in this study were in the same range as rate constants for anaerobic HCH biodegradation determined in laboratory degradation experiments (Langenhoff et al., 2001; Quintero et al., 2005).

Pollutant	Flow path	Distance [m]	Residence time [d]	Year	B$^\Delta$ [%]	λ_s [1/m]	s$_{1/2}$ [m]	λ_t [1/d]	t$_{1/2}$ [d]
α-HCH	A→B	52	175	2010	40	0.0098	71	0.0029	235
β-HCH	A→B	52	175	2008	34*	0.0080	86	0.0024	287
				2010	86	0.0370	19	0.0111	62
δ-HCH	A→B	52	175	2008	16*	0.0032	214	0.0010	715
				2010	19*	0.0041	171	0.0012	569
	A→C	98	327	2008	30*	0.0037	190	0,0011	632
				2010	50	0.0070	99	0.0021	331
	E→F	71	237	2008	67	0.0156	44	0.0047	148
				2010	75	0.0192	36	0.0058	120

$^\Delta$ Biodegradation was calculated on the basis of previously published anaerobic γ-HCH fractionation (ε= -3.9 ‰ (Badea et al., 2009)).
* Biodegradation is not significant as δ^{13}C values used for calculation differ by less than ± 2 ‰.

Table 1: Percentage of biodegradation (B [%]), distance-dependent and time-dependent *in situ* first-order biodegradation rate constants (λ_s [1/m], λ_t [1/d]) as well as half-life distances and times (s$_{1/2}$ [m], t$_{1/2}$ [d]) for HCHs calculated for flow paths of the investigated transect in 2008 and 2010.

Conclusions

Carbon isotope ratios of HCHs in combination with pollutant concentration patterns and field site background information were used in this study to demonstrate for the first time:
- *in situ* measurement of HCH biodegradation in the field,
- quantification of HCH biodegradation within different zones of the contaminant plume,
- CSIA applicability for the assessment of biodegradation and source identification of HCHs in the environment,
- Time-resolved CSIA can reveal temporal variations in HCH biodegradation and provide information on the influences of various processes on natural attenuation.

Due to the intensive production of HCHs and their worldwide usage, there are a huge number of HCH-contaminated production, formulation, and dump sites (Vijgen et al., 2011). At these sites, time-resolved CSIA could be instrumental to identify trends in pollutant attenuation and help to predict the evolution of contaminant plumes, as exemplified in this study. *In situ* biodegradation rate constants could be integral in modeling the current status and future development of contaminant plumes. Thus, CSIA possesses the potential for improved prediction of HCH distribution at contaminated field sites. These outcomes suggest that CSIA constitutes a viable monitoring tool and could be beneficial for the implementation and control of innovative management and remediation concepts like *Monitored* or *Enhanced Natural Attenuation* (MNA, ENA).

Acknowledgments

We gratefully acknowledge Ursula Günther, Falk Bratfisch and Silviu Badea for their analytical support with isotope analysis. We also thank the water authority, Consultant Company and owner of the field site for sampling and providing chemical and hydrological data as well as historical information. The University of Agriculture Faisalabad, Pakistan and HIGRADE are acknowledged for the financial support of Safdar Bashir.

Supporting Information

Additional information on standard procedures for groundwater sampling and concentration analyses, CSIA of HCHs, sum concentration of HCHs in 2008, hydrogeochemical parameters, and qualitative assessment of degradation for α-, β- and γ-HCH. This material is available in Chapter A Appendix.

References

Aelion C. M., Höhener, P., Hunkeler, D., & Aravena, R. (Eds.). (2010).Environmental isotopes in biodegradation and bioremediation, 1st edn. CRC Press, Boca Raton.

Alvarez A., Benimeli C.S., Saez J.M., Fuentes M.S., Cuozzo S.A., Polti M.A., Amoroso M.J. (2012) Bacterial bio-resources for remediation of Hexachlorocyclohexane. International Journal of Molecular Sciences, 13(11):15086-15106.

Badea S.L., Vogt C., Weber S., Danet A.F., Richnow H.H. (2009) Stable isotope fractionation of γ-Hexachlorocyclohexane (Lindane) during reductive dechlorination by two strains of sulfate-reducing bacteria. Environmental Science & Technology, 43(9):3155-3161.

Badea S.L., Vogt C., Gehre M., Fischer A., Danet A.F., Richnow H.H. (2011) Development of an enantiomer-specific stable carbon isotope analysis (ESIA) method for assessing the fate of α-hexachlorocyclohexane in the environment. Rapid Commun. Mass Spectrom, 25(10):1363-1372.

Bashir S., Fischer A., Nijenhuis I., Richnow H.H. (2013) Enantioselective carbon stable isotope fractionation of Hexachlorocyclohexane during aerobic biodegradation by *Sphingobium* spp. Environmental Science & Technology, 47(20):11432-11439.

Bhatt P., Kumar M.S., Chakrabarti T. (2009) Fate and degradation of POP-hexachlorocyclohexane. Critical Reviews in Environmental Science Technology, 39(8):655-695.

Bombach P., Richnow H.H., Kastner M., Fischer A. (2010) Current approaches for the assessment of *in situ* biodegradation. Applied Microbiology and Biotechnology, 86(3):839-852.

Breivik K., Pacyna J.M., Münch J. (1999) Use of α-, β-and γ-hexachlorocyclohexane in Europe, 1970–1996. Science of the Total Environment, 239(1):151-163.

Cichocka D., Imfeld G., Richnow H.H., Nijenhuis I. (2008) Variability in microbial carbon isotope fractionation of tetra- and trichloroethene upon reductive dechlorination. Chemosphere, 71(4):639-648.

Coplen T.B. (2011) Guidelines and recommended terms for expression of stable-isotope-ratio and gas-ratio measurement results. Rapid Communication in Mass Spectrometry, 25(17):2538-2560.

Coplen T.B., Brand W.A., Gehre M., Groning M., Meijer H.A.J., Toman B., Verkouteren R.M. (2006) New guidelines for delta C-13 measurements. Analytical Chemistry, 78(7):2439-2441.

EEA (2007) Progress in management of contaminated sites (CSI 015) - Assessment published Aug 2007 - Progress in management of contaminated sites.

Elsner M. (2010) Stable isotope fractionation to investigate natural transformation mechanisms of organic contaminants: principles, prospects and limitations. Journal of Environmental Monitoring, 12(11):2005-2031.

Hatzinger P.B., Bohlke J.K., Sturchio N.C. (2013) Application of stable isotope ratio analysis for biodegradation monitoring in groundwater. Current Opinion in Biotechnology, 24(3):542-549.

Heinisch E., Kettrup A., Bergheim W., Martens D., Wenzel S. (2005) Persistent chlorinated hydrocarbons (PCHC), source oriented monitoring in aquatic media - 3. The isomers of hexachlorocyclohexane. Fresenius Environmental Bulletin, 14(6):444-462.

Hofstetter T.B., Berg M. (2011) Assessing transformation processes of organic contaminants by compound-specific stable isotope analysis. Trac-Trends in Analytical Chemistry, 30(4):618-627.

Hunkeler D., Meckenstock, R.U., Sherwood-Lollar, B., Schmidt, T., Wilson, J.T. (2008) A Guide for Assessing Biodegradation and Source Identification of Organic Groundwater Contaminant using Compound Specific Isotope Analysis. 600/R-08/148; U.S. Environmental Protection Agency: Washington, DC, 2008.

Illman W.A., Alvarez P.J. (2009) Performance assessment of bioremediation and natural attenuation. Critical Reviews in Environmental Science & Technology, 39(4):209-270.

Jagnow G., Haider K., Ellwardt P.C.H.R. (1977) Anaerobic dechlorination and degradation of hexachlorocyclohexane isomers by anaerobic and facultative anaerobic bacteria. Archives of Microbiology, 115(3):285-292.

Lal R., Pandey G., Sharma P., Kumari K., Malhotra S., Pandey R., Raina V., Kohler H.P.E., Holliger C., Jackson C., Oakeshott J.G. (2010) Biochemistry of microbial degradation of hexachlorocyclohexane and prospects for bioremediation. Microbiology and Molecular Biology Review, 74(1):58-80.

Langenhoff A.A.M., Staps S.J.M., Pijls C., Rijnaarts H.H.M. (2013) Stimulation of Hexachlorocyclohexane (HCH) biodegradation in a full scale *in situ* bioscreen. Environmental Science & Technology, 47(19):11182-11188.

Langenhoff A.A.M., Staps J.J.M., Pijls C., Alphenaar A., Zwiep G., Rijnaarts H.H.M. (2001) Intrinsic and stimulated *in situ* biodegradation of Hexachlorocyclohexane (HCH). 6th International HCH and Pesticides forum book:181-186.

Li S.L., Elliott D.W., Spear S.T., Ma L.M., Zhang W.X. (2011) Hexachlorocyclohexanes in the environment: Mechanisms of dechlorination. Critical Reviews in Environmental Science & Technology, 41(19):1747-1792.

Li Y. (1999) Global technical hexachlorocyclohexane usage and its contamination consequences in the environment: from 1948 to 1997. Science of the Total Environment, 232(3):121-158.

Lotse E.G., Graetz D.A., Chesters G., Lee G.B., Newland L.W. (1968) Lindane adsorption by lake sediments. Environmental Science & Technology, 2(5):353-357.

Mariotti A., Germon J.C., Hubert P., Kaiser P., Letolle R., Tardieux A., Tardieux P. (1981) Experimental determination of nitrogen kinetic isotope fractionation: some principles; illustration for the denitrification and nitrification processes. Plant Soil, 62(3):413-430.

Meckenstock R.U., Morasch B., Griebler C., Richnow H.H. (2004) Stable isotope fractionation analysis as a tool to monitor biodegradation in contaminated aquifers. Journal of Contaminant Hydrology, 75(3-4):215-255.

Mehboob F., Langenhoff A.A.M., Schraa G., Stams A.J.M. (2013) Anaerobic degradation of Lindane and other HCH isomers, in: A. Malik, et al. (Eds.), Management of microbial resources in the environment, Springer, Dordrecht. pp. 495-521.

Minh N.H., Minh T.B., Kajiwara N., Kunisue T., Subramanian A., Iwata H., Tana T.S., Baburajendran R., Karuppiah S., Viet P.H., Tuyen B.C., Tanabe S. (2006) Contamination by persistent organic pollutants in dumping sites of Asian developing countries: Implication of emerging pollution sources. Archives of Environmental Contamination and Toxicology, 50(4):474-481.

Phillips T.M., Lee H., Trevors J.T., Seech A.G. (2006) Full-scale *in situ* bioremediation of hexachlorocyclohexane-contaminated soil. Journal of Chemical Technology and Biotechnology, 81(3):289-298.

Quintero J.C., Moreira M.T., Feijoo G., Lema J.M. (2005) Anaerobic degradation of hexachlorocyclohexane isomers in liquid and soil slurry systems. Chemosphere, 61(4):528-536.

Ritter L., Solomon K., Sibley P., Hall K., Keen P., Mattu G., Linton B. (2002) Sources, pathways, and relative risks of contaminants in surface water and groundwater: A perspective prepared for the Walkerton inquiry. Journal of Toxicology and Environmental Health-Part a-Current Issues, 65(1):1-142.

Sahsuvar L., Helm P.A., Jantunen L.M., Bidleman T.F. (2003) Henry's law constants for α-, β-, and γ-hexachlorocyclohexanes (HCHs) as a function of temperature and revised estimates of gas exchange in Arctic regions. Atmosphere Environment, 37(7):983-992.

Schirmer M., Dahmke A., Dietrich P., Dietze M., Gädeke S., Richnow H.H., Schirmer K., Weiß H., Teutsch G. (2006) Natural attenuation research at the contaminated megasite Zeitz. Journal of Hydrology, 328(3-4):393-407.

Schwarzenbach R.P., Egli T., Hofstetter T.B., von Gunten U., Wehrli B. (2010) Global water pollution and human health. Annual Review of Environment and Resources, 35:109-136.

Soga K., Page J.W.E., Illangasekare T.H. (2004) A review of NAPL source zone remediation efficiency and the mass flux approach. Journal of Hazardous Material, 110(1-3):13-27.

Thullner M., Centler F., Richnow H.H., Fischer A. (2012) Quantification of organic pollutant degradation in contaminated aquifers using compound-specific stable isotope analysis - review of recent developments. Organic Geochemistry, 42(12):1440-1460.

Vijgen J., Abhilash P., Li Y.F., Lal R., Forter M., Torres J., Singh N., Yunus M., Tian C., Schäffer A. (2011) Hexachlorocyclohexane (HCH) as new Stockholm Convention POPs - a global perspective on the management of Lindane and its waste isomers. Environmental Science and Pollution Research, 18(2):152-162.

Walker K., Vallero D.A., Lewis R.G. (1999) Factors influencing the distribution of Lindane and other hexachlorocyclohexanes in the environment. Environmental Science & Technology, 33(24):4373-4378.

Weber R., Watson A., Forter M., Oliaei F. (2011) Persistent organic pollutants and landfills - a review of past experiences and future challenges. Waste Management & Research, 29(1):107-121.

Weber R., Gaus C., Tysklind M., Johnston P., Forter M., Hollert H., Heinisch E., Holoubek I., Lloyd-Smith M., Masunaga S., Moccarelli P., Santillo D., Seike N., Symons R., Torres J.P.M., Verta M., Varbelow G., Vijgen J., Watson A., Costner P., Woelz J., Wycisk P., Zennegg M. (2008) Dioxin- and POP-contaminated sites-contemporary and future relevance and challenges. Environmental Science and Pollution Research, 15(5):363-393.

Wiedemeier T.H., Rifai H.S., Newell C.J., Wilson J.T. (1999) Natural attenuation of fuels and chlorinated solvents in the subsurface. 1 ed. John Wiley & Sons, Inc., New York.

Zoumis T., Schmidt A., Grigorova L., Calmano W. (2001) Contaminants in sediments: remobilisation and demobilisation. Science of the Total Environment, 266(1-3):195-202.

Chapter : 4 Discussion

Discussion

Discussion

Carbon stable isotope analysis (CSIA) has emerged as an efficient tool to detect, characterize, and to quantify biodegradation of organic and inorganic contaminants in the environment. An important feature of this approach is that it allows destructive losses of contaminants to be distinguished from losses caused by non-destructive processes such as dilution, dispersion, and sorption (Hatzinger et al., 2013). Concentration dependent monitoring approaches cannot explain well if the changes in concentration of contaminant are due to destructive processes or non-destructive processes. The most common application of stable isotope analysis for monitoring contaminants fate is to identify the incidence of transformation based on changes in a single isotope ratio (e.g. $^{13}C/^{12}C$). When a contaminant exhibits spatial or temporal variations in concentration, the isotope ratios may indicate whether this is caused by biodegradation or by other processes with different patterns of isotopic variation (e.g. source changes) or by non-fractionating processes (e.g. dilution). Under favorable circumstances, the magnitude of isotope effects (variations in isotope ratios) can be related to the progress of the biodegradation reaction where concentration changes alone may be difficult to interpret (Hatzinger et al., 2013). The isotopic approach has been used to document degradation of many common organic groundwater contaminants, including chlorinated solvents (Meckenstock et al., 2004; Wiedemeier, 1999; Wijker et al., 2013) and main degradation intermediates of these solvents (Hunkeler, 2009). Thus it is important to understand how specific environmental conditions influence the changes in isotope ratios. Presence of extremely high concentrations of HCH across the world as described in the general introduction chapter, need efficient tools for monitoring HCH natural attenuation. In this context, this book provides insight into the applicability and development of CSIA as a tool to monitor *in situ* transformation of HCH by providing various evidences with findings from laboratory and field investigations.

CSIA and characterization of HCH transformation

For applying CSIA in the course of biodegradation quantification, it is necessary to choose the right enrichment factor (ε_c) representative for the prevailing biogeochemical degradation processes at the examined field site (Wiedemeier, 1999). This indicates that information on the biochemical factors affecting the *in situ* biodegradation including isotope fractionation is available particularly to assess the uncertainty of the CSIA approach. To gain reliable data, intensive studies on the variability of enrichment factors, typical for certain environmental conditions have to be performed prior to CSIA application for quantification of *in situ* transformation. In case of HCH, a first study was done by Badea and colleagues (2009) by applying CSIA to monitor changes in stable carbon isotope ratios during biodegradation of γ-HCH under sulfate reducing conditions. They could successfully validate the applicability of CSIA for HCH transformation and provided the basis for the application of isotope techniques *in situ*, by providing carbon enrichment factors (ε_c) for anoxic environments. α-HCH is a racemate and its enantio-selective transformation may provide a second clue for transformation and can be used as second indicator to confirm transformation of HCH (instead of multi element stable isotope analysis). In a subsequent study, Badea and colleagues developed the analytical method to analyze the changes in isotopic composition of individual enantiomers of α-HCH and proposed the enantiomer-specific stable isotope analysis (ESIA) concept for the anaerobic biotransformation with *Clostridium pasteurianum* (Badea et al., 2011). This provided the basis to apply this approach for assessing *in situ* transformation not only for HCH but also for other chiral contaminants.

In this book, CSIA was used as primary tool to characterize transformation of HCH isomers in various reference experiments, which include biotic **(Chapter 3.1 & 3.2)** and abiotic **(Chapter 3.3)** transformation processes of HCH. The objective of this book was to validate the applicability of CSIA as an effective tool to characterize transformation of HCH under various environmental conditions and this work provides a step further to validate the applicability of CSIA for environmental applications. The first study, in this regard, was investigation of changes in isotope ratios of HCH isomers and enantiomers during aerobic transformation **(Chapter 3.1)**. For the reliable quantification of *in situ* degradation an enrichment factor representative for environmental conditions is needed. Thus reference cultures isolated from typical contaminated environments should be used to obtain fractionation factors. So two aerobic bacterial strains were selected (*S. indicum* strain B90A and *S. japonicum* strain UT26) which were commonly found in aerobic HCH contaminated sites (Lal et al., 2010). The estimated stable carbon isotope enrichment factors for γ-HCH transformation determined in this study were the same for both aerobic reference bacterial strains with $\varepsilon_c = -1.5 \pm 0.1$ ‰ and -1.7 ± 0.2 ‰ for *S. indicum* strain B90A and *S. japonicum* strain UT26, respectively **(Chapter 3.1)**. Thus, similar reaction mechanisms for γ-HCH biodegradation can be expected for both strains, which was supported by the similarity of γ-HCH *lin* genes coding the *lin* protein in both strains (Lal et al., 2010). The ε_c values calculated for aerobic bacterial strains for γ-HCH showed smaller carbon isotope fractionation as compared to those obtained for various anaerobic reference experiments (ε_c = −5.5 ± 0.8 ‰ for *D. mccartyi* strain 195 , ε_c = −4.1 ± 0.6 ‰ for *Clostridium pasteurianum* DSMZ 525 **(Chapter 3.2)**. The anaerobic enrichment factors calculated in our study were similar in range as calculated previously (−3.6 ‰ to −3.9 ‰) (Badea et al., 2009). These difference in enrichment factors quantified under oxic and anoxic conditions confirmed the different reaction mechanisms which were reported previously as anaerobic γ-HCH biodegradation is initiated by reductive beta-elimination (Quintero et al., 2005) and aerobic degradation by dehydrochlorination (Lal et al., 2010). Thus the extent and range of carbon enrichment factors determined in our studies prove that CSIA can be used to characterize various degradation pathways of HCH.

Badea and colleagues reported carbon enrichment factors for γ-HCH under sulfate reducing conditions for anoxic environment. However, it is also important to understand the variability of enrichment factors or changes in isotope ratios due to other *in situ* processes such as variation in metabolic or co-metabolic transformation of substrate, microbial competition in mixed cultures and presence of mixed isomers. In this context, studies with various reference culture experiments considering anaerobic processes showed that there were no significant differences among enrichment factors determined with pure cultures which metabolically degrade γ-HCH (*D. mccartyi* strain 195 , ε_c = −5.5 ± 0.8 ‰), co-metabolically degrading pure cultures (*C. pasteurianum* DSMZ 525, ε_c = −4.1 ± 0.6 ‰) and by field enrichment cultures, ε_c = −3.1 ± 0.4 ‰. Reference experiments with mixed HCH isomer showed similar enrichment factor for γ-HCH as determined in anaerobic biodegradation experiments with γ-HCH alone **(Chapter 3.2)**. Analysis of the carbon stable isotope composition during anaerobic biotransformation confirmed the similarity in degradation pathways under anoxic conditions which were confirmed with similar metabolite patterns **(Chapter 3.2)**. This suggests that the isotope sensitive steps under anaerobic conditions have a similar mechanism upon bond cleavage. These reference experiments showed the consistency of enrichment factors for quantifying the *in situ* biodegradation of γ-HCH in anoxic environments.

In case of α-HCH isomers, similar carbon isotope enrichment factors were estimated for *S. indicum* strain B90A ($\varepsilon_c = -1.6 \pm 0.3$ ‰) and for *S. japonicum* strain UT26 (-1.0 ± 0.2 ‰). The ε_c values of α-HCH showed no substantial differences when compared with those of γ-HCH. This indicates that both HCH isomers were transformed by similar reaction mechanism with both bacterial strains as reported previously (Lal et al., 2010). However, ε_c values for aerobic cultures suggested smaller isotope fractionation than for anaerobic degradation of α-HCH ($\varepsilon_c = -3.7 \pm 0.8$ ‰) (Badea et al., 2011). This may indicate that different reaction mechanisms under oxic and anoxic environment lead to different modes of bond cleavage in the initial reaction of the transformation pathway (Lal et al., 2010). However, since aerobic biodegradation was significantly faster than anaerobic degradation, steps other than isotope sensitive carbon bond cleavage might be rate limiting for the overall aerobic biodegradation of α-HCH and therefore can mask the carbon isotope fractionation (Aeppli et al., 2009; Nijenhuis et al., 2005). Future studies using multi element isotope analysis may characterize the mode of C-Cl or C-H bond cleavage and may provide more mechanistic insight. For this study methods for chlorine and hydrogen isotope fractionation for HCH were not available.

For assessing potential masking effects during biotic isotopic fractionation reference experiments, relevant abiotic fractionation experiments were performed **(Chapter 3.3)**. Alkaline hydrolysis was assumed as similar reaction to aerobic dehydrochlorination reaction because of the production of a similar initial metabolite (Trantirek et al., 2001). Higher isotope fractionation was noticed for α-HCH (ε_c bulk = -7.6 ± 0.4 ‰) during abiotic alkaline hydrolysis as compared to aerobic transformation. Thus non-isotope fractionating steps (e.g., substrate uptake into the cell, binding of the substrate to the enzyme) during biodegradation might be more rate-limiting as compared to direct bond cleavage in abiotic reaction. In future CSIA studies with pure enzymes and two dimentional compound specific stable isotope analysis might explain better the biotic transformation process of HCH and can help further to validate the assumptions for rate limitations during isotope sensitive steps during aerobic biodegradation processes.

Fe(0) reduction and anaerobic biodegradation give almost identical fractionation factors. Comparing anaerobic biodegradation and relevant abiotic transformation reaction by nano iron, there was no significant difference in calculated enrichment factors **(Chapter 3.3)**. The metabolite pattern of anaerobic degradation suggests a mechanism similar to dichloroelimination by two-electron transfer during Fe^0 reduction. Taking Fe^0 reduction as a model reaction for anaerobic biodegradation the isotope sensitive steps were not significantly masked by potential rate limitation by uptake or enzyme binding of the HCH.

Under oxic conditions, experiments were conducted at different temperatures to assess the masking effects due to changes in growth conditions (for example, temperature can affect biodegradation rates). Experiments to elucidate the dependency of isotope fractionation for α-HCH degradation by *S. indicum* strain B90A at optimal temperature condition at 30°C, 20°C and 10°C led to slower biodegradation rates but did not change isotope fractionation significantly. This suggests that overall degradation kinetics does not affect carbon isotope fractionation in the temperature range between 10 and 30°C. So, the relative smaller carbon isotope fractionation for aerobic HCH biodegradation seems to be mainly caused by the reaction mechanism and the carbon isotope enrichment factor is representative for assessing HCH biodegradation under typical temperature conditions in oxic environments **(Chapter 3.1)**.

Enantiomer fractionation (EF) and assessing *in situ* transformation of HCH

Enantiomer fractionation (EF) provides another indicator for *in situ* biodegradation of α-HCH according to previous observations (Harner et al., 2000; Helm et al., 2000; Law et al., 2004). Considering the concept of enantiomer-specific stable isotope analysis (ESIA) and EF, various studies were performed to validate their applicability for *in situ* transformation of the chiral α-HCH. The enantioselective transformation of chiral isomers may also give a clue to assess *in situ* transformation of α-HCH. In aerobic degradation experiments (-) α-HCH was preferentially degraded in both reference culture experiments **(Chapter 3.1)**. The preferential transformation of (-) α-HCH has been previously reported in soil (Falconer et al., 1997; Finizio et al., 1998). Soil samples represent an aerobic environment so similar microbial processes can be assumed as observed in our reference experiments with aerobic bacterial strains which are frequently found in HCH contaminated sites (Lal et al., 2010). ESIA combined with EF and CSIA may thus be used to trace aerobic degradation using the enrichment factors from our reference fractionation experiments.

In case of all abiotic reference experiments, non enantioselective transformation of α-HCH was observed, as expected for chemical reactions **(Chapter 3.3)**. The enantioselective transformation is solely dependent on microbial process as no enantioselective reactions were observed in a previous study with sterilized sludge (Mueller and Buser, 1995). Enantiomer specific biodegradation may be due to differences in the preferential enzymatic activity of the enantiomers for the aerobic biotransformation process. In this case, the *lin* genes (genes expressed during HCH transformation in aerobic bacterial strains) necessary for aerobic degradation of HCH may be taken in to account. These *lin* genes were initially identified and characterized for *S. japonicum* strain UT26 and were subsequently recovered from *S. indicum* strain B90A as well (Lal et al., 2010) which suggests that similar enzymes are involved in catalysis of α-HCH. However, *S. japonicum* strain UT26 contains LinA while *S. indicum* strain B90A is known to express two copies of *linA* (*linA1* and *linA2*) (Dogra et al., 2004). The amino acid sequences of the products encoded by the *linA1* and *linA2* genes are 88 % matching to each other and 88% (LinA1) and 99 % (LinA2) are similar to the sequence of LinA of *S. japonicum* UT26 (Imai et al., 1991). Preferential degradation of (-) α- enantiomer is known with the LinA1 variant of *S. indicum* B90A (Suar et al., 2005) which was also confirmed by our study **(Chapter 3.1)**. By applying the Rayleigh equation, we obtained the enantiomer enrichment factors for aerobic α-HCH biodegradation **(Chapter 3.1)**, providing a quantitative framework which can be determined by common GC analysis and it is similar to CSIA approach. Moreover, this approach might also be applicable for other enantiomeric pollutants as it was applied for O-desmethylvenlafaxine (Gasser et al., 2012; Maier et al., 2013).

The abiotic reference experiments did not show enantioselective transformation and this information can be used to distinguish *in situ* biotic and abiotic transformation **(Chapter 3.3)**. This has to be taken critically as anaerobic experiments performed by Badea and colleagues also showed non-enantioselective transformation of α-HCH by *C. pasterianum*, which poses uncertainty when identifying reactions under anoxic conditions using ESIA. For evaluation of this issue more studies are needed to confirm the validity of this approach for comparing anaerobic and abiotic transformation.

Enantiomer-specific carbon isotope fractionation

Significant carbon isotope fractionation was observed in our studies with aerobic reference culture experiments, which was different for each of the α-HCH

Chapter 4: Discussion

enantiomers. Since each individual enantiomer may have different rates of degradation, quantification of *in situ* biodegradation based on enantiomer-specific isotope analysis allows more refined assessment of contaminated field sites with chiral pollutants, as it was confirmed by a recent field study (Milosevic et al., 2012). This enantiomer-specific analysis is especially important only if one enantiomer is subject to biodegradation while the other persists. In our study we found variability in the carbon enrichment factors (ε_c) for each enantiomer in case of ESIA for aerobic transformations **(Chapter 3.1)** but abiotic studies showed no differences in carbon isotope enrichment factors (ε_c) for each enantiomer **(Chapter 3.3)**. Thus various biotic and abiotic processes can be characterized *in situ* by using ESIA in combination with EF. ESIA is proposed to be an additional tool to monitor *in situ* transformation of other chiral pesticides and pharmaceuticals in future.

The combination of enantiomer and carbon isotope fractionation, in a two-dimensional approach, using values of EF(+) and EF(-) in correlation with the stable carbon isotope discrimination ($\Delta\delta^{13}C$) of the α-HCH enantiomers may allow to trace degradation pathways of α-HCH in the environment **(Chapter 3.1 and Chapter 3.3)**. The specific ranges for the pathways obtained in our study can be used as reference for evaluating field data in order to determine the relevance of biotic and abiotic transformation of chiral contaminants.

The combined isotope and enantiomeric fractionation has the potential to differentiate biotic and abiotic transformation of α-HCH. In order to combine both approaches, EF(+) was plotted against the carbon isotope discrimination ($\Delta = \delta^{13}C_0 - \delta^{13}C_t$) of the (+) α-enantiomer as well as EF(-) against the carbon isotope discrimination of (-) α-enantiomer **(Chapter 3.1 and Chapter 3.3)** and compared to previous investigation on anaerobic α-HCH biodegradation (Badea et al., 2011). These different trends indicate the potential for distinguishing biotic and abiotic transformation of α-HCH in the environment based on changes in isotope ratios ($\delta^{13}C$) and enantiomer fractionation (EF). Due to lack of anaerobic enantioselective biodegradation studies anaerobic transformation cannot be fully distinguished from abiotic transformation as there is overlap also in the extent of carbon isotope fractionation. Therefore, more anaerobic enantiomer selective studies are needed to validate the model for distinguishing anaerobic and abiotic transformation employing enantiomer fraction and enantioselective stable isotope fractionation.

Dehalococcoides sp. and in situ HCH transformation

One of the specific scientific gaps for assessing *in situ* transformation of HCH, using molecular biological techniques, is the lack of well characterized microbial cultures. Strains of the genus *Dehalococcoides* have received significant attention because of their ability to reductively dehalogenate common groundwater contaminants such as chlorinated ethenes, ethanes and benzenes, due to their extensive genomic inventory of putative reductive dehalogenases, mostly with unknown function (Kaufhold et al., 2013; Krajmalnik-Brown et al., 2007; Löffler et al., 2013; Pöritz et al., 2013; Seshadri et al., 2005). Kaufhold, *et al.*, (2013) reported the transformation of γ-HCH isomer with a highly enriched *D. mccartyi* strain BTF08-containing culture, as well as by *D. mccartyi* strain 195 (Kaufhold et al., 2013). However it did not show the growth dependent dechlorination process, or the biotransformation of other HCH isomers. In **Chapter 3.2** for the first time, the metabolic biotransformation of γ-HCH was shown for an isolate, *D. mccartyi* strain 195 which used hydrogen as electron donor and γ-HCH as terminal electron acceptor. Degradation rates of different isomers were in the following order γ-HCH > α-HCH > β-HCH=δ-HCH which are in similar order as reported previously for co-metabolic reactions (Jagnow et al., 1977). Mixed cultures enriched from contaminated field site also show the similar trends in HCH isomers transformation γ-HCH > α-HCH > β-HCH > δ-HCH. The extent and range of the changes in carbon stable isotope ratios of γ-HCH during biotransformation confirmed the similarity in degradation pathways under anoxic conditions and suggests that carbon stable isotope analysis can be used as means to quantify the *in situ* biodegradation of γ-HCH under anoxic conditions.

Characterization of abiotic transformation α-HCH using CSIA

Possible abiotic mechanisms for α-HCH transformation in the environment include photo-induced reactions, hydrolysis, electrochemical reduction and reduction by Fe^0 nanoparticles. It is also important to understand if CSIA can characterize these environmentally relevant abiotic processes. In this book all selected reference studies could clearly show the potential of CSIA to characterize abiotic transformations of α-HCH.

Enrichment factor determined for UV/H_2O_2 ($\varepsilon_c = -1.9 \pm 0.2$ ‰) process was similar to those of from aerobic biodegradation of α-HCH with *S. indicum* strain B90A ($\varepsilon_c = -1.6 \pm 0.3$ ‰) and *S. japonicum* strain UT26 ($\varepsilon_c = -1.0 \pm 0.2$ ‰), respectively **(Chapter**

3.1), whereas direct photolysis, electrochemical and Fe^0 reduction revealed the same magnitude of isotope enrichment factor (average ε_c = -3.7 ± 0.8 ‰) as observed in anaerobic biodegradation of α-HCH with *C. pasterianum* (Badea et al., 2011). The isotope enrichment factors observed for the UV/H_2O_2 process had a similar magnitude as the one for aerobic biodegradation by a dehydrochlorination, which resulted pentachlorocyclohexene (PCCH) as the first intermediate which is subsequently converted to 1,2,4-TCB (Butler and Hayes, 1998). Dehydrochlorination of α-HCH by alkaline hydrolysis and aerobic biodegradation resulted in PCCH and TCBs as metabolites, which may indicate that similar reaction mechanisms or transformation pathways are taking place. However, when comparing the enrichment factors of alkaline hydrolysis (ε_{bulk} = −7.6 ± 0.4 ‰) and aerobic biodegradation with *S. indicum* strain (ε_{bulk} = −1.6 ± 0.3 ‰, and $\varepsilon_{\alpha\text{-}HCH}$ = −1.0 ± 0.2 ‰) clear differences were observed **(Chapter3.1)**. Although both transformation reactions might be similar, as they produce similar metabolites, the chemical mechanism of bond cleavage might be different as reflected in the carbon isotope fractionation pattern.

In addition, it is expected that the binding step of the chemical to the enzyme during the aerobic biodegradation is to some extent rate limiting, which might contribute to the lower isotope fractionation as compared to the pure chemical reaction of alkaline hydrolysis in a homogeneous solution. This rate limitation can mask the extent of isotope fractionation for biotic transformation (Northrop, 1981). Substrate uptake into the cell and binding of the substrate to the enzyme can reduce the isotope effect compared to direct bond cleavage in a chemical reaction (Aeppli et al., 2009). Conclusions on the mechanism of biological dehydrochlorination in analogy to alkaline hydrolysis are difficult to obtain based on metabolite pattern and extent of isotope fractionation.

The differences in enrichment factors between abiotic **(Chapter 3.3)** and aerobic transformation **(Chapter 3.1)** process indicate the applicability of CSIA for the characterization α-HCH in both environmental conditions. The overlap in fractionation factors indicates difficulties to distinguish the characterization of α-HCH transformation under abiotic and anaerobic environment. In future multi element isotope analysis may be employed for characterizing difference between abiotic and anaerobic reaction mechanisms.

CSIA as tool to assess *in situ* transformation

In this book, a novel approach was developed for monitoring *in situ* biodegradation incorporating the isotope signatures of isomers and enantiomers of HCH as well as enantiomer selective degradation. The isotope enrichment factors for the transformation of HCH under biotic (α- and γ-HCH) and abiotic (α-HCH) conditions were estimated for the first time allowing the quantification of HCH biodegradation in different environmental compartments. Additionally, the differences in isotope fractionation between aerobic and anaerobic degradation indicates that CSIA can be used to evaluate degradation pathways and can be applied to monitor natural attenuation of HCH.

To apply these laboratory findings to real field conditions a field monitoring campaign was performed to assess the potential of carbon compound-specific stable isotope analysis for the assessment of biodegradation and source identification within a contaminated aquifer at a former pesticide processing facility **(Chapter 3.4)**. Hydrogeochemical conditions of the contaminated aquifer showed anaerobic environment in two sampling campaigns in 2008 and 2010. The wells selected for sampling were along the flow direction of ground water with the order of downstream from A to F **(Chapter 3.4)**. Both years and in all wells, δ-HCH was found to be the most abundant and concentrations of the other HCH isomers were at least three times smaller (Being β-HCH more abundant than α- and γ-HCH). The concentration patterns of HCHs at the investigated field site revealed a similar trend of biodegradability as observed in many degradation experiments: δ-HCH≈β-HCH >α-HCH >γ-HCH (Quintero et al., 2005). The low concentration of α- and γ-HCH and presence of δ and β-HCH may be due to preferential degradation of one isomer over the other occurring in the field **(Chapter 3.4)**.

The observed concentration decrease from 2008 to 2010 was accompanied by ^{13}C-enrichment as observed in biodegradation fractionation reference experiments **(Chapter 3.1 & 3.2)**. Conservative calculations based on the Rayleigh equation exhibited varying levels of HCH biodegradation (16-86 %) in investigated aquifer. Moreover, temporal and spatial *in situ* first-order biodegradation rate constants were estimated with maximal values of 0.0029 d^{-1} and 0.0098 m^{-1} for α-HCH, 0.0110 d^{-1} and 0.0370 m^{-1} for β-HCH, and 0.0058 d^{-1} and 0.0192 m^{-1} for δ-HCH, respectively **(Chapter 3.4)**. The biodegradation rate constants (λ_t –values) obtained in this study were in the same range as rate constants for anaerobic HCH biodegradation determined in laboratory degradation experiments (Langenhoff et al., 2013; Quintero et al., 2005). Thus, it could be assumed that δ-HCH was the most recalcitrant HCH isomer with respect to calculated biodegradation rate constants and half-life values. A high recalcitrance of δ-HCH under anoxic conditions was observed in our mixed cultures laboratory reference experiment **(Chapter 3.2)** and it was also suggested in other studies (Jagnow et al., 1977; Quintero et al., 2005).

Another application of CSIA is to assess sources of contamination in complex contaminated aquifers (Meckenstock et al., 2004) and it has been applied successfully in different studies (Nijenhuis et al., 2013). The carbon isotope ratios in conjunction with concentration of HCHs revealed two main contaminant source zones located at wells A and D/E. The HCH source at well A resulted to be from a contamination at the former processing facilities and the source at wells D and E from the former dumping of HCH wastes **(Chapter 3.4)**. Carbon isotope ratios of HCHs provided evidence of HCH biodegradation within the groundwater downstream of the main HCH source zones at wells C and F, indicating that biodegradation contributed to the natural attenuation of HCHs within the investigated aquifer **(Chapter 3.4)**. From 2008 to 2010, concentrations of HCHs decreased concomitant with changes in carbon isotope ratios at most of the wells indicating that the contribution of biodegradation for the natural attenuation of HCHs increased.

Thus due to the intensive production of HCHs and their worldwide usage, there is a huge number of HCH contaminated production, formulation and dump sites (Vijgen et al., 2011). At these sites, time-resolved CSIA might be applied in order to identify trends in pollutant removal and help to predict the evolution of contaminant plumes which is shown in this study. As suggested by this study, CSIA would be an appropriate monitoring tool and would be beneficial for the implementation and successful control of innovative management and remediation concepts like Monitored or Enhanced Natural Attenuation (MNA, ENA).

References

Aeppli C., Berg M., Cirpka O.A., Holliger C., Schwarzenbach R.P., Hofstetter T.B. (2009) Influence of mass-transfer limitations on carbon isotope fractionation during microbial dechlorination of trichloroethene. Environmental Science & Technology, 43(23):8813-8820.

Badea S.L., Vogt C., Weber S., Danet A.F., Richnow H.H. (2009) Stable isotope fractionation of γ-Hexachlorocyclohexane (Lindane) during reductive dechlorination by two strains of sulfate reducing bacteria. Environmental Science & Technology, 43(9):3155-316.

Badea S.L., Vogt C., Gehre M., Fischer A., Danet A.F., Richnow H.H. (2011) Development of an enantiomer-specific stable carbon isotope analysis (ESIA) method for assessing the fate of α-hexachlorocyclohexane in the environment. Rapid Communications in Mass Spectrometry, 25(10):1363-1372.

Butler E.C., Hayes K.F. (1998) Effects of solution composition and pH on the reductive dechlorination of Hexachloroethane by iron sulfide. Environmental Science & Technology, 32(9):1276-1284.

Dogra C., Raina V., Pal R., Suar M., Lal S., Gartemann K.H., Holliger C., van der Meer J.R., Lal R. (2004) Organization of lin genes and IS6100 among different strains of hexachlorocyclohexane-degrading *Sphingomonas paucimobilis*: evidence for horizontal gene transfer. Journal of Bacteriology, 186(8):2225-2235.

Falconer R.L., Bidleman T.F., Szeto S.Y. (1997) Chiral pesticides in soils of the Fraser Valley, British Columbia. Journal of Agricultural and Food Chemistry, 45(5):1946-1951.

Finizio A., Bidleman T., Szeto S. (1998) Emission of chiral pesticides from an agricultural soil in the Fraser Valley, British Columbia. Chemosphere, 36(2):345-355.

Gasser G., Pankratov I., Elhanany S., Werner P., Gun J., Gelman F., Lev O. (2012) Field and laboratory studies of the fate and enantiomeric enrichment of venlafaxine and O-desmethylvenlafaxine under aerobic and anaerobic conditions. Chemosphere, 88(1):98-105.

Harner T., Jantunen L.M., Bidleman T.F., Barrie L.A., Kylin H., Strachan W.M., Macdonald R.W. (2000) Microbial degradation is a key elimination pathway of hexachlorocyclohexanes from the Arctic Ocean. Geophysical Research Letters, 27(8):1155-1158.

Hatzinger P.B., Böhlke J., Sturchio N.C. (2013) Application of stable isotope ratio analysis for biodegradation monitoring in groundwater. Current Opinion in Biotechnology, 24(3):542-549.

Helm P.A., Diamond M.L., Semkin R., Bidleman T.F. (2000) Degradation as a loss mechanism in the fate of α-hexachlorocyclohexane in Arctic watersheds. Environmental Science & Technology, 34(5):812-818.

Hunkeler D., Meckenstock, R.U., Sherwood-Lollar, B., Schmidt, T., Wilson, J.T. (2009) A Guide for Assessing Biodegradation and Source Identification of Organic Groundwater Contaminant using Compound Specific Isotope Analysis (CSIA). 600/R-08/148; U.S. Environmental Protection Agency: Washington, DC.

Imai R., Nagata Y., Fukuda M., Takagi M., Yano K. (1991) Molecular cloning of a Pseudomonas paucimobilis gene encoding a 17-kilodalton polypeptide that eliminates HCl molecules from γ-hexachlorocyclohexane. Journal of Bacteriology, 173(21):6811-6819.

Jagnow G., Haider K., Ellwardt P.C.H.R. (1977) Anaerobic dechlorination and degradation of hexachlorocyclohexane isomers by anaerobic and facultative anaerobic bacteria. Archives of Microbiology, 115(3):285-292

Kaufhold T., Schmidt M., Cichocka D., Nikolausz M., Nijenhuis I. (2013) Dehalogenation of diverse halogenated substrates by a highly enriched Dehalococcoides-containing culture derived from the contaminated mega-site in Bitterfeld. FEMS Microbiology Ecology, 83(1):176-188.

Krajmalnik-Brown R., Sung Y., Ritalahti K.M., Saunders F.M., Loffler F.E. (2007) Environmental distribution of the trichloroethene reductive dehalogenase gene (tceA) suggests lateral gene transfer among *Dehalococcoides*. FEMS Microbiology Ecology, 59(1):206-214.

Lal R., Pandey G., Sharma P., Kumari K., Malhotra S., Pandey R., Raina V., Kohler H.P.E., Holliger C., Jackson C., Oakeshott J.G. (2010) Biochemistry of Microbial Degradation of Hexachlorocyclohexane and prospects for bioremediation. Microbiology and Molecular Biology Reviews, 74(1):58-80.

Langenhoff A.A., Staps S.J., Pijls C., Rijnaarts H.H. (2013) Stimulation of Hexachlorocyclohexane (HCH) Biodegradation in a full Scale *in situ* bioscreen. Environmental Science & Technology, 47(19):11182-11188.

Law S.A., Bidleman T.F., Martin M.J., Ruby M.V. (2004) Evidence of enantioselective degradation of α-hexachlorocyclohexane in groundwater. Environmental Science & Technology, 38(6):1633-1638.

Löffler F.E., Ritalahti K.M., Zinder S.H. (2013) *Dehalococcoides* and reductive dechlorination of chlorinated solvents, Bioaugmentation for Groundwater Remediation, Springer. pp. 39-88.

Maier M.P., Qiu S., Elsner M. (2013) Enantioselective stable isotope analysis (ESIA) of polar herbicides. Analytical and Bioanalytical Chemistry, 405(9):2825-2831.

Meckenstock R.U., Morasch B., Griebler C., Richnow H.H. (2004) Stable isotope fractionation analysis as a tool to monitor biodegradation in contaminated acquifers. Journal of Contaminant Hydrology, 75(3):215-255.

Milosevic N., Qiu S., Elsner M., Einsiedl F., Maier M., Bensch H., Albrechtsen H.-J., Bjerg P.L. (2012) Combined isotope and enantiomer analysis to assess the fate of phenoxy acids in a heterogeneous geologic setting at an old landfill. Water Research, 47(2):637–649.

Mueller M.D., Buser H.-R. (1995) Environmental behavior of acetamide pesticide stereoisomers. 2. Stereo-and enantioselective degradation in sewage sludge and soil. Environmental Science & Technology, 29(8):2031-2037.

Nijenhuis I., Andert J., Beck K., Kästner M., Diekert G., Richnow H.-H. (2005) Stable isotope fractionation of tetrachloroethene during reductive dechlorination by *Sulfurospirillum multivorans* and *Desulfitobacterium* sp. strain PCE-S and abiotic reactions with cyanocobalamin. Applied and Environmental Microbiology, 71(7):3413-3419.

Nijenhuis I., Schmidt M., Pellegatti E., Paramatti E., Richnow H.H., Gargini A. (2013) A stable isotope approach for source apportionment of chlorinated ethene plumes at a complex multi-contamination events urban site. Journal of Contaminant Hydrology, 153:92-105.

Northrop D.B. (1981) The expression of isotope effects on enzyme-catalyzed reactions. Annual Review of Biochemistry, 50(1):103-131.

Pöritz M., Goris T., Wubet T., Tarkka M.T., Buscot F., Nijenhuis I., Lechner U., Adrian L. (2013) Genome sequences of two dehalogenation specialists– *Dehalococcoides mccartyi* strains BTF08 and DCMB5 enriched from the highly polluted Bitterfeld region. FEMS Microbiology Letters, 343(2):101-104.

Quintero J.C., Moreira M.T., Feijoo G., Lema J.M. (2005) Anaerobic degradation of hexachlorocyclohexane isomers in liquid and soil slurry systems. Chemosphere, 61(4):528-536.

Seshadri R., Adrian L., Fouts D.E., Eisen J.A., Phillippy A.M., Methe B.A., Ward N.L., Nelson W.C., Deboy R.T., Khouri H.M. (2005) Genome sequence of the PCE-dechlorinating bacterium *Dehalococcoides ethenogenes*. Science, 307(5706):105-108.

Suar M., Hauser A., Poiger T., Buser H.R., Müller M.D., Dogra C., Raina V., Holliger C., van der Meer J.R., Lal R. (2005) Enantioselective transformation of α-hexachlorocyclohexane by the dehydrochlorinases LinA1 and LinA2 from the soil bacterium *Sphingomonas paucimobilis* B90A. Applied and Environmental Microbiology, 71(12):8514-8518.

Trantirek L., Hynkova K., Nagata Y., Murzin A., Ansorgova A., Sklenar V., Damborsky J. (2001) Reaction mechanism and stereochemistry of γ-hexachlorocyclohexane dehydrochlorinase LinA. Journal of Biological Chemistry, 276:7734-40.

Vijgen J., Abhilash P., Li Y.F., Lal R., Forter M., Torres J., Singh N., Yunus M., Tian C., Schäffer A. (2011) Hexachlorocyclohexane (HCH) as new Stockholm Convention POPs-a global perspective on the management of Lindane and its waste isomers. Environmental Science and Pollution Research, 18(2):152-162.

Wiedemeier T.H. (1999) Natural attenuation of fuels and chlorinated solvents in the subsurface. John Wiley & Sons.

Wijker R.S., Bolotin J., Nishino S.F., Spain J.C., Hofstetter T.B. (2013) Using compound-specific isotope analysis to assess biodegradation of nitroaromatic explosives in the subsurface. Environmental Science & Technology, 47(13):6872-6883

Chapter: 5 Summary & outlook

Summary & outlook

Summary & outlook

The objective of this book, therefore, was to validate the application of carbon stable isotope analysis (CSIA) and enantiomer specific stable isotope analysis (ESIA) to characterize biotic and abiotic transformation of HCH *in situ*. Additionally, to prove that CSIA in combination with enantiomer-selective degradation of α-HCH can be applied as an effective and reliable tool for monitoring natural attenuation of HCH.

The reaction-specific isotope enrichment factors (ε_c) were determined in laboratory experiments for α- and γ -HCH isomers during aerobic **(Chapter 3.1)** and anaerobic **(Chapter 3.2)** degradation and compared with relevant abiotic reactions **(Chapter 3.3)**. Bulk enrichment factors (ε_c) determined for aerobic degradation of α and γ-HCH by two *Sphingobium* spp. with similar reaction mechanism were similar for γ-HCH (ε_c = -1.5 ± 0.1 ‰ and -1.7 ± 0.2 ‰) and α-HCH (ε_c = -1.0 ± 0.2 ‰ and -1.6 ± 0.3 ‰). The enrichment factors calculated for oxic transformation of γ-HCH are smaller as compared to determined under anoxic γ-HCH transformation with the ε_c = -5.5 ± 0.8 ‰ for *D. mccartyi* strain 195 (metabolically degrading), ε_c = -3.1 ± 0.4 ‰ for the enrichment culture and ε_c = -4.1 ± 0.6 ‰ for *C. pasteurianum* DSMZ 525 (co-metabolically degrading). This difference is explained by different reaction mechanisms taking place during aerobic and anaerobic conditions. Furthermore, isomer and enantiomers selective stable isotope fractionation of chiral isomer α-HCH was estimated during biotic **(Chapter 3.1)** and abiotic **(Chapter 3.3)** reactions. Enrichment factors (ε_c) calculated with biotic reactions were smaller as resulted in same abiotic reaction (dehydrochlorination), considering the metabolites patterns, whereas direct photolysis, electrochemical and Fe^0 reduction revealed the same magnitude of isotope enrichment factors as observed in anaerobic biodegradation **(Chapter 3.2., 3.3)**. This can be explained with limiting effects during biotic degradation such as substrate uptake into the cell, binding of the substrate to the enzyme as compared to direct bond cleavage in a chemical reaction. Therefore, carbon isotope enrichment factors ε_{bulk} derived from our laboratory findings may help to understand the differences between biotic and abiotic reactions taking place in field studies but to confirm its applicability 2D isotope analysis (C, Cl) may help. Non enantioselective transformation and fractionation of α-HCH was observed in all abiotic transformation reactions **(Chapter 3.3)** as compared to enantioselective degradation and fractionation in biotic transformation **(Chapter 3.1)**. The enrichment factors of individual enantiomers $\varepsilon_{enantiomer}$ allows to calculate an average factor in all cases which was identical with bulk enrichment factors ε_{bulk} showing the validity of the analytical approach. The variability in enrichment factors observed for enantiomers ($\varepsilon_{enantiomer}$) during biotic investigation showed the preferential reactivity of enzymes for one enantiomer over the other which is not observed in abiotic studies. Thus CSIA combined with ESIA and enantiomer fractions (EF) can help to distinguish biotic and abiotic reactions taking place *in situ*.

The calculated stable carbon isotope enrichment factors ε_c allowed the estimation of the extent of HCH degradation at the contaminated field site **(Chapter 3.4)**. Application of CSIA led to comparable estimates of biodegradation and contaminant source identification using carbon isotope ratios. So, the data bank of enrichment factors provided by this study can help to quantify the biodegradation *in situ*. These studies show that CSIA is useful technique to quantify *in situ* degradation of HCH and identify reaction pathways, by combining ESIA and (EF).

In short, this study provides a concept allowing the use of enantiomer fractionation, CSIA and ESIA for a complementary and thus comprehensive assessment of *in situ* degradation of α-HCH, but also other chiral contaminants. Thus, this book offers proof of principle for the theoretical framework of a multiple lines of evidence approach by i) providing evidence of degradation, ii) distinguishing pathways and iii) quantifying HCH degradation at contaminated field sites.

However, as other degradation pathways such as dehydrogenation and oxidation may be responsible for HCH biodegradation in a field sites in addition to the so far known reductive beta-elimination or dehydrogenation, reference culture experiments with pure cultures having these respective degradation pathways should be performed. Carbon isotope enrichment factors ε_{bulk} derived from this study help us to understand the differences between biotic and abiotic. The overlap in fractionation factors indicates difficulties to distinguish the characterization of α-HCH transformation under abiotic and anaerobic environment. In future multi element isotope analysis may be employed for characterizing difference between abiotic and anaerobic reaction mechanisms.

There is also urgent need to study in-depth molecular and biochemical basis of HCH transformation under anaerobic environment. Growth dependent potential of *Dehalococcoides* sp. to degrade HCH have been proved in this study **(Chapter 3.2)** but need to be confirmed if it has same potential for HCH isomers. Studies regarding expression of proteins and enzymes and experiments with pure enzymes can explain the variability of enrichment factors calculated under different reference experiments i.e. biotic and abiotic. More anaerobic reference culture experiments are needed to differentiate abiotic and anaerobic enantioselective transformation of α-HCH.

Additionally, the applicability of the combined enantiomer fractionation, CSIA and ESIA approach needs to be tested in field studies. Also field monitoring campaigns for assessing the applicability of CSIA for aerobic environment will authenticate the CSIA techniques developed in this book.

Appendix

Supporting Information for Chapter 3.1

Enantioselective carbon stable isotope fractionation of hexachlorocyclohexane during aerobic biodegradation by *Sphingobium* spp.

Safdar Bashir[1], Anko Fischer[1,2], Ivonne Nijenhuis[1], and Hans-Hermann Richnow[1]

[1]Department of Isotope Biogeochemistry, Helmholtz Centre for Environmental Research – UFZ, Permoserstraße 15, 04318, Leipzig, Germany.
[2]Isodetect - Company for Isotope Monitoring, Permoserstr. 15, 04318 Leipzig, Germany

A1: Supporting information chapter 3.1

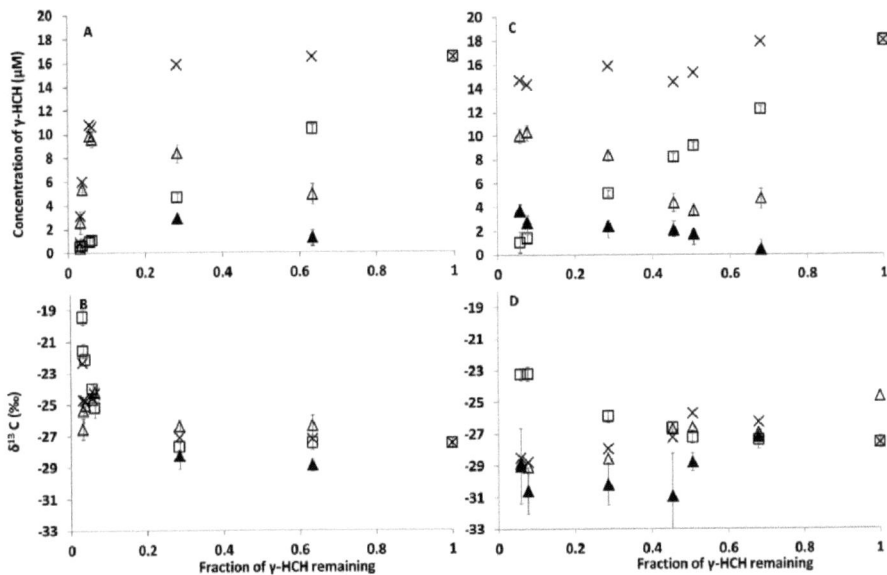

Figure S1. Biodegradation of γ-HCH by S. indicum strain B90A and S. japonicum strain UT26: Concentrations and carbon isotope ratios of γ-HCH and its metabolites. Concentrations (A) and carbon isotope ratios (B) of γ-HCH and products during biodegradation by S. indicum strain B90A. γ-HCH (open square) and its metabolites 1,2,4-TCB (open triangle) and γ-pentachlorocyclohexene (γ-PCCH) (close triangle). Concentrations (C) and carbon isotope ratios (D) of γ-HCH and products during biodegradation by S. japonicum strain UT26. γ-HCH (open square) and its metabolites 1,2,4-TCB (open triangle) and 2,5-dichlorophenol (2,5-DCP) (close triangle). The sum concentration (A,C) and average carbon isotope composition (B, D) of substrate and products (see Materials and Methods for calculation) are indicated by X. Error bars indicate the standard deviation of triplicate analysis for carbon isotope analysis.

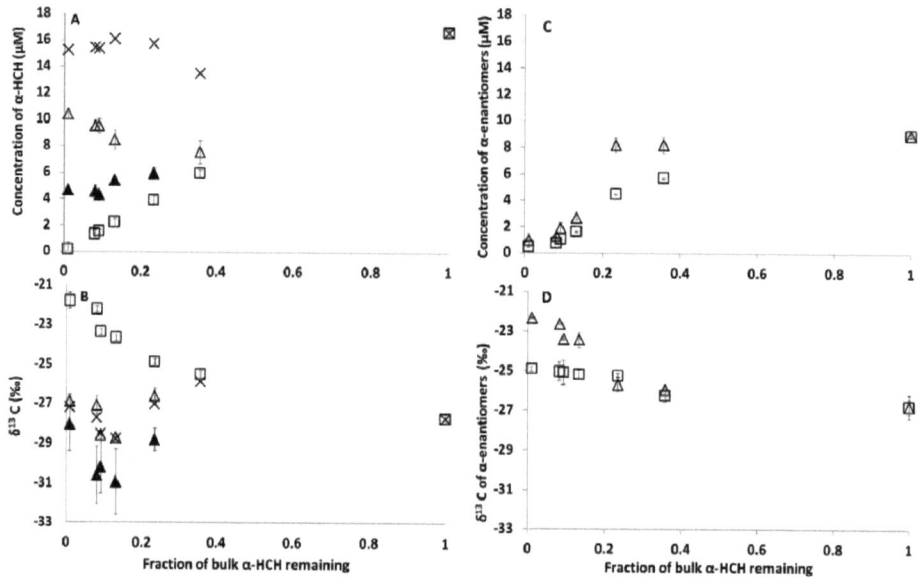

Figure S2. Biodegradation of α-HCH by S. japonicum strain UT26: Concentrations and carbon isotope ratios of α-HCH, its enantiomers and metabolites.

Concentrations (A) and carbon isotope ratios (B) of α-HCH and metabolites during biodegradation by S. japonicum strain UT26. α-HCH (open square) and metabolites 1,2,4-TCB (open triangle) and 2,5-dichlorophenol (2,5-DCP) (close triangle). Changes in concentrations (C) and carbon isotope ratios (D) of (-) α-HCH (open square) and (+) α-HCH (open triangle) during biodegradation by S. japonicum strain UT26. The sum concentration (A) and average carbon isotope composition (B) of substrate and products are indicated by X. Error bars indicate the standard deviation of triplicate analysis for carbon isotope analysis.

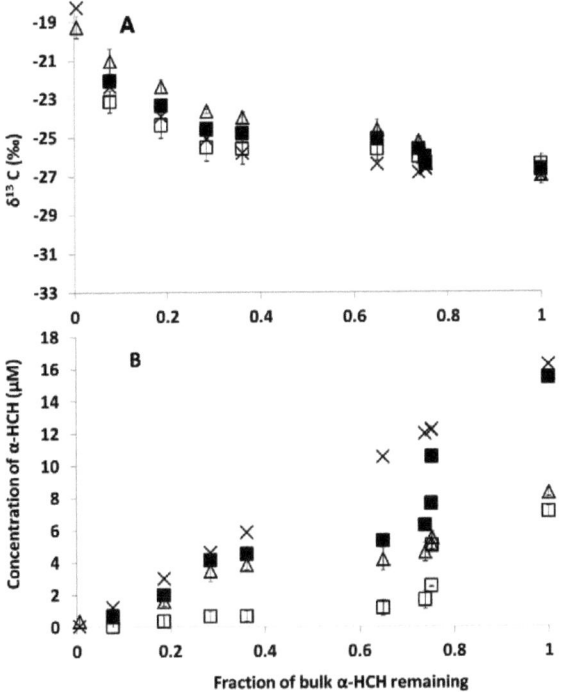

Figure S3. Comparison of average concentrations (A) and average carbon isotope composition (B) of α-HCH and its enantiomers during biodegradation by S. indicum strain B90A.

Bulk α-HCH directly analyzed (cross) and bulk α-HCH calculated by using (+) α-HCH and (-) α-HCH enantiomer concentrations and carbon isotope ratios (closed squares). (-) α-HCH (open square) and (+) α-HCH (open triangle).

Supporting Information for Chapter 3.2

Anaerobic biotransformation of hexachlorocyclohexane isomers by *Dehalococcoides* spp.

Safdar Bashir, Kevin Kuntze, Carsten Vogt and Ivonne Nijenhuis

Department of Isotope Biogeochemistry, Helmholtz Centre for Environmental Research – UFZ, Permoserstraße 15, 04318, Leipzig, Germany.

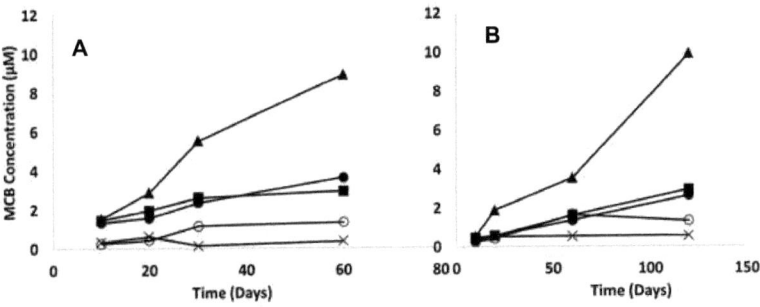

Figure S1: Production of MCB as putative end-product in the transformation of α- (filled circles), β- (open circles), γ- (filled triangles) and δ-HCH (filled squares) by *D. mccartyi* strain 195 (A) and *D. mccartyi* strain BTF08 (B). Degradation experiments were done with a single dose of each isomer and each point represents the average of triplicate bottles. The standard deviation at each point was < 5%. Crosses represent the average concentration of MCB in controls without inoculumas average for all the tested isomers.

Table S1: HCH isomers recovered after incubation with *Dehalococcoides* strains

Isomer	*Dehalococcoides mccartyi* strain 195 (60 days incubation)	*Dehalococcoides* ribotype BTF08 (120 days incubation)
	% HCH isomer recovered approx.	% HCH isomer recovered approx.
α-HCH	86	90
β-HCH	98	90
γ-HCH	65	43
δ-HCH	95	96

Supporting Information for Chapter 3.3

Compound Specific Stable Isotope Analysis (CSIA) to Characterize Abiotic Reaction Mechanisms of Alpha-Hexachlorocyclohexane

Ning Zhang[1], Safdar Bashir[1], Jinyi Qin[2], Anko Fischer[1, 3], Ivonne Nijenhuis[1], Hartmut Herrmann[4], Lukas Y. Wick[2], and Hans H. Richnow[1]

[1]Department of Isotope Biogeochemistry, Helmholtz Centre for Environmental Research-UFZ, Permoserstraße 15, 04318 Leipzig, Germany

[2]Department of Environmental Microbiology, Helmholtz Centre for Environmental Research-UFZ, Permoserstraße 15, 04318 Leipzig, Germany

[3]Isodetect – Company for Isotope Monitoring, Deutscher Platz 5B, 04103 Leipzig, Germany

[4]Department of Chemistry, Leibniz Institute for Tropospheric Research (TROPOS), Chemistry Dept., Permoserstraße 15, 04318 Leipzig, Germany

Chemicals

All chemicals were of analytical grade and used without further purification. Alpha-hexachlorocyclohexane (α-HCH, analytical purity, 99%), hexachlorobenzene (HCB), ferric chloride anhydrous (FeCl$_3$), sodium borohydride (NaBH$_4$) were purchased from Sigma-Aldrich. Hydrogen peroxide (H$_2$O$_2$) (30% w/w), sodium bicarbonate (Na$_2$CO$_3$), sodium carbonate (NaHCO$_3$) and n-pentane (analytical purity >99%) were supplied by Merck (Darmstadt, Germany). The water used for preparing solutions was ultrapure water (obtained by Milli-Q System, Millipore GmbH, Schwalbach/Ts. Germany).

Preparation of α-HCH stock solution

Because of its low solubility in water (10 ppm) (Clayton et al., 1991), α-HCH was dissolved in 10 mL acetone at a concentration of 1000 mg L^{-1} in the presence of silica gel (10 mg). Ultra sonication (0.5 h) was used for adsorption of α-HCH on the silica gel surface. Then acetone was completely evaporated at 60°C (0.5 h). The silica gel was added into aqueous phosphate buffer (pH 7.0, 10 mM) and the mixture was shaken overnight for desorption of α-HCH. The concentration of α-HCH in this phosphate buffer solution was 1 mg L^{-1}. This α-HCH stock solution was used in all experiments described in this study, except alkaline hydrolysis.

Setup for the photochemical transformation experiments

A schematic diagram of the photochemical reactor system was shown in Figure S1. The reactor was a 200 mL Pyrex cylindrical flask with a quartz window whose surface area was approximately 28 cm^2. Irradiation was achieved by applying a 150 W xenon lamp (Type No.: L2175, Hamamatsu, Japan), which covered a broad continuous spectrum from the ultraviolet to infrared region (185-2000 nm). For direct photolysis at $\lambda \geq 185$ nm, the light source was directly faced to the quartz window of the reactor. A filter with a cut-off wavelength $\geq$ 280 nm was applied in the case of indirect photolysis. The temperature was controlled at 20°C by a circulating water system surrounding the Pyrex reactor. The reaction solution was magnetically stirred during the whole experiment.

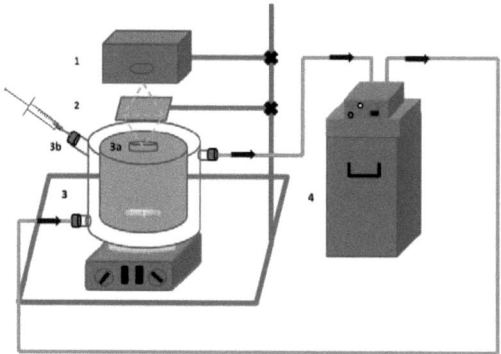

Figure S1. The setup for the photochemical experiments. 1. 150 W xenon lamp; 2. filter with a 280 nm cut-on wavelength; 3. reactor: 3a.quartz window; 3b. sampling ports; 4. cooling water system.

Setup for the electrochemical transformation experiment

Two inert Ti/Ir coated grids (13 cm × 3.2 cm) were inserted as electrodes into a stirred 500 mL beaker, which was filled with 300 mL of the α-HCH stock solution and covered with a Teflon membrane to prevent evaporation. An electric field strength of 6.3 V cm^{-1} (1.68 mA cm^{-2}) was applied.

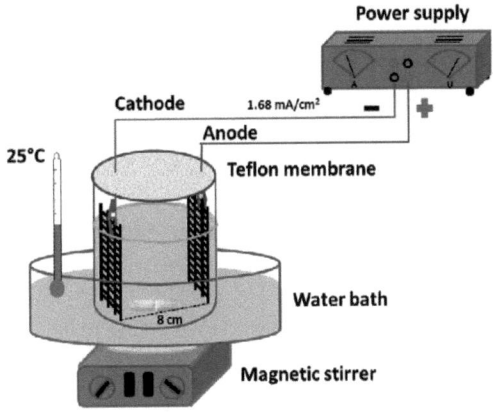

Figure S2. The setup for the electrochemical experiment.

Analytical methods

The concentration of α-HCH was determined by gas chromatograph equipped with a flame ionization detector (GC-FID, 7820A, Agilent Technologies, USA). A HP-5 column (30 m length × 0.32 mm ID, 0.25 μm film thickness; J & W Scientific, Agilent Technologies, USA) was used for separation. The temperature program started at 40°C for 5 min, and then increased to 180°C at a rate of 10°C min^{-1} held for 2 min, and finally increased to 300°C with the rate of 20°C min^{-1} held for 5 min. A volume of 2 μL of extract was injected in split mode with a split ratio 1:1 into a hot injector held at 250°C. The concentration of α-HCH was calculated by using the response factor, which is defined as the ratio of sensitivities of the analyte and the internal standard (Patnaik, 2011). Monitoring of corresponding main reaction products was done by gas chromatograph-mass spectrometry (GC-MS) as described elsewhere (Bashir et al., 2013). A BPX-5 capillary column (30 m length × 0.25 mm ID × 0.25 μm film thickness; SGE, Darmstadt, Germany) was used for separation. The oven temperature program was the same as described for the GC-FID analysis. Carbon stable isotope ratios of α-HCH were measured by a gas chromatograph-combustion-isotope ratio mass spectrometer (GC-C-IRMS) as described in previous work (Bashir et al., 2013) and expressed in the delta notation ($\delta^{13}C$) according to eq. S1 in SI. A Zebron ZB1 column (60 m length ×

0.32 mm ID × 1 μm film thickness; Phenomenex, Aschaffenburg, Germany) was used for separation. For chiral α-HCH analysis, a Gamma DEX™ 120 chiral column was applied (30 m length × 0.25 mm ID × 0.25 μm film thickness; Supelco Bellefonte, PA) on which minus (−) and plus (+) α-HCH could be separated. Each sample was analyzed in triplicate and the precision of $\delta^{13}C$-values for α-HCH and its enantiomers was ≤ 0.5‰.

Quantification of carbon isotope fractionation

The carbon isotope ratios were expressed in the delta notation ($\delta^{13}C$) relative to the international standard Vienna Pee Dee Belemnite (V-PDB) according to eq. (S1) (Coplen, 2011):

$$\delta^{13}C_{sample} = \frac{R_{sample}}{R_{stanard}} - 1 \quad (S1)$$

In equation (S1), R_{sample} and $R_{standard}$ were the $^{13}C/^{12}C$ ratios of the sample and V-PDB, respectively. Because variations in carbon isotope abundance was typically small, $\delta^{13}C$-values were reported in part per thousand (‰ or per mil).

For the description of stabile isotope fractionation of chemical reactions the Rayleigh equation can be applied (Elsner, 2010; Mariotti et al., 1981):

$$\frac{(\delta_t+1)}{(\delta_0+1)} = \left(\frac{C_t}{C_0}\right)^{\varepsilon_C} \quad (S2)$$

where δ_t, δ_0 and C_t, C_0 were the stable carbon isotope ratios and concentrations of a compound at a given point in time (t) and at the beginning of a transformation reaction (0), respectively. The isotope enrichment factor (ε_C) correlated the changes in stable isotope ratios [$(\delta_t+1)/(\delta_0+1)$] with the changes in the concentrations (C_t/C_0).

The isotope enrichment factor (ε_C) can be determined from the logarithmic form of the Rayleigh equation:

$$\ln\left[\frac{(\delta_t+1)}{(\delta_0+1)}\right] = \varepsilon_C \ln\left(\frac{C_t}{C_0}\right) \quad (S3)$$

plotting $\ln(C_t/C_0)$ versus $\ln[(\delta_t+1)/(\delta_0+1)]$ and obtaining from the slope of the linear regression m = $\varepsilon_C/1000$. Since carbon isotope enrichment factors were typically small, ε_C-values were reported in part per thousand (‰). The error of the isotope enrichment factors were given as 95% confidence interval (CI) determined by a regression curve analysis (Elsner et al., 2007).

For quantification of the kinetic isotope effect (KIE) characterizing the bond cleavage, the isotope enrichment factor (ε_C) need to be normalized to the isotope effect of the bond cleavage. The calculation of apparent kinetic isotope effect (AKIE) allowed comparing isotope effects of bond changes, which can be used to gain insights into the chemical reaction mechanisms. AKIE were calculated according to (Elsner et al., 2005):

$$\mathrm{AKIE}_C = \left(\frac{1}{z \times \frac{n}{x} \times \varepsilon_C + 1} \right) \quad (S4)$$

where n was the number of atoms of the molecule of a selected element, x was the number of reactive positions, and z was the number of positions in intra-molecular competition. The AKIE_C quantified the isotope effect of the bond change using experimental data. The uncertainty of the AKIE_C (err. AKIE_C) was estimated by error propagation (eq. S5) by taking into account the error of the isotope enrichment factor (err. ε_C):

$$\mathrm{err.\ AKIE}_C = \left| \frac{\partial \mathrm{AKIE}_C}{\partial \varepsilon} \right| \times \mathrm{err.\ } \varepsilon_C \quad (S5)$$

Photochemical reactions

UV absorption spectrum. The resulting spectrum of the α-HCH stock solution (oxygen free) was shown in Figure S3. The measurement was performed under oxygen-free conditions by flushing the UV spectrometer with argon. A purged phosphate buffer (pH 7.0, 10 mM) was used as the reference. The maximum of absorbance was observed at λ = 252 nm, which was in agreement with the reported high absorption at λ = 255 nm ($\varepsilon_{255\ nm}$ = 1000 M^{-1} cm^{-1}) (Fiedler et al., 1993).

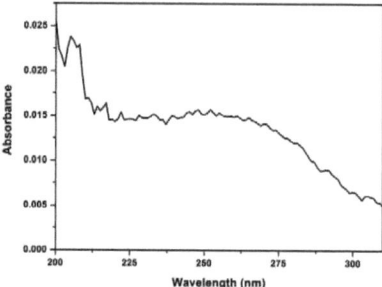

Figure S3. UV absorption spectrum of the α-HCH stock solution purged with argon.

Control experiments. For photochemical reactions, control experiments without light (only with H_2O_2) and only with UV irradiation ($\geq$ 280 nm) showed almost stable concentration and suggested that the sole presence of H_2O_2 and direct photolysis in longer wavelengths ($\geq$ 280 nm) did not affect concentration of α-HCH under the experimental conditions used.

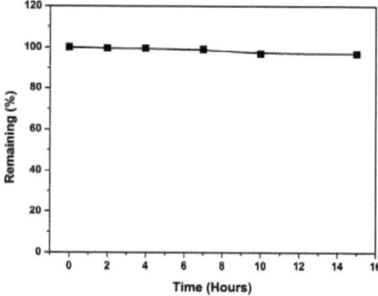

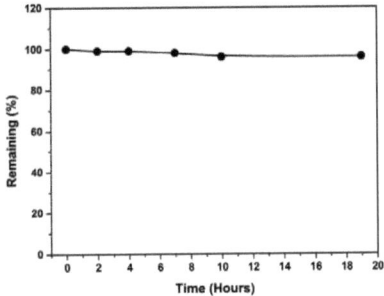

Figure S4. Remaining fraction of α-HCH in the control experiments without UV irradiation only with H_2O_2 (square) and under UV irradiation at wavelength ≥ 280 nm (circle) in phosphate buffer (pH=7.0).

Product analysis and transformation pathways. The chemical structures of transformation products were analyzed by GC-MS in order to obtain information on the mechanisms of the chemical reactions investigated in this study.

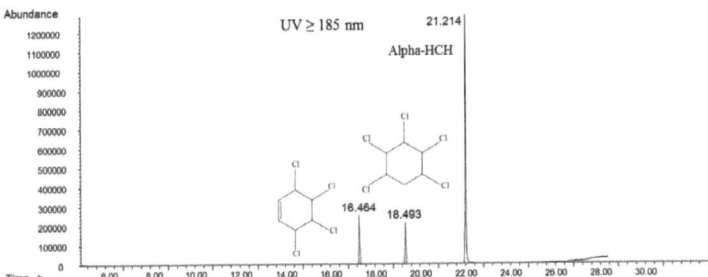

Figure S5. Total ion chromatogram of transformation products from direct photolysis upon radiation with the UV wavelength ≥ 185 nm.

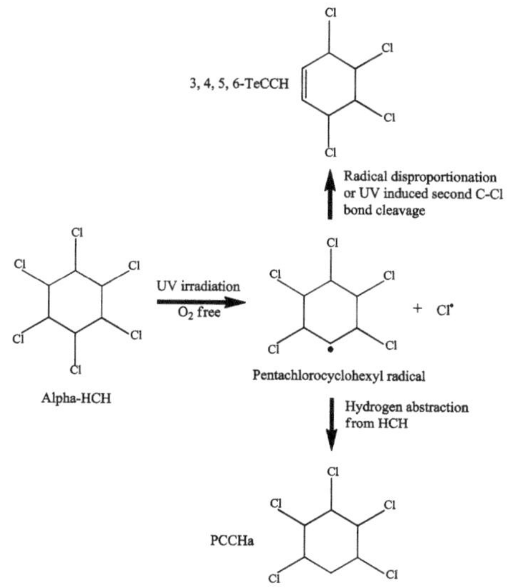

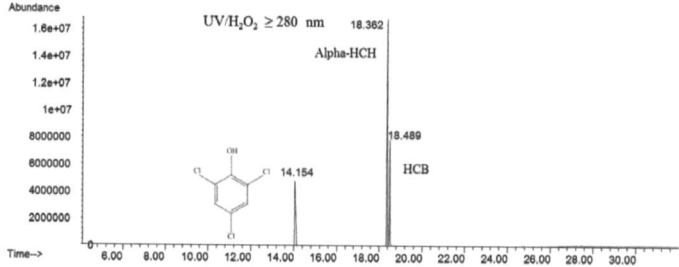

Scheme S1. Proposed reaction mechanism for α-HCH during direct photolysis in the UV wavelength ≥ 185 nm.

Figure S6. Total ion chromatogram of transformation products of UV/H$_2$O$_2$ reaction using irradiation in the range of ≥ 280 nm. HCB was used as internal standard.

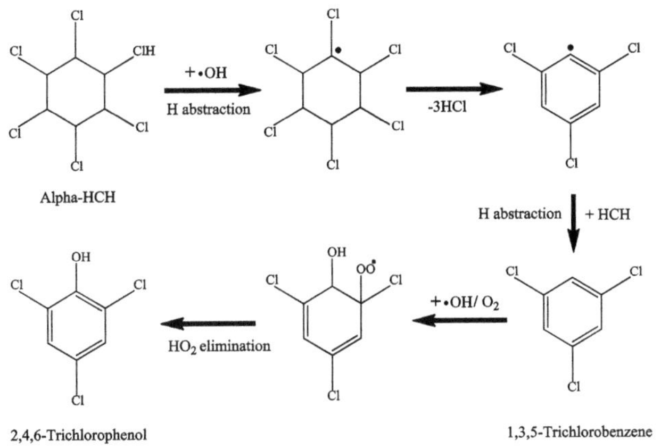

Scheme S2. Proposed reaction mechanism for α-HCH via UV/H$_2$O$_2$ process in the UV range ≥ 280 nm.

Alkaline hydrolysis

Control experiment. As we described, the stock solution of α-HCH was prepared and stored in the pH 7 phosphate buffered solution and used for each reaction except alkaline hydrolysis (prepared in Na$_2$CO$_3$-NaHCO$_3$ buffer, pH 9.78). As a control experiment, the concentration of α-HCH in the stock solution was measured every two days within the monitoring period of one month. The concentration stayed stable, indicating no transformation or evaporation loss of α-HCH took place in this neutral buffered solution (data not shown).

Product analysis and transformation pathways. The GC-MS analysis revealed that 1,3,4,5,6-pentachlorocyclohexene (C$_6$H$_4$Cl$_5$, PCCH), 1,2,4-trichlorobenzene (1,2,4-TCB) and 1,2,3-trichlorobenzene (1,2,3-TCB) were formed during alkaline hydrolysis (Figure S7). The transformation products were in agreement with the proposed pathway of dehydrochlorination from γ-HCH to PCCH and TCBs (Liu et al., 2003).

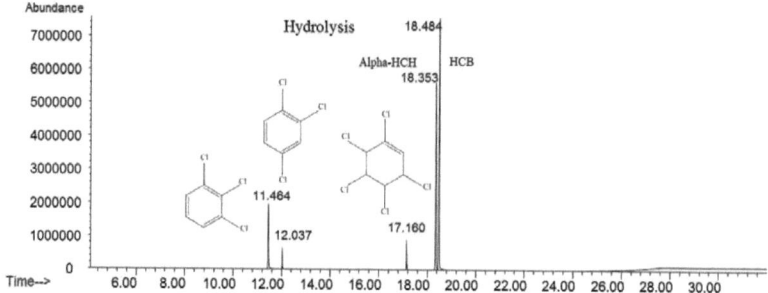

Figure S7. Total ion chromatogram of transformation products of alkaline hydrolysis at pH 9.78 in Na_2CO_3-$NaHCO_3$ buffer. HCB was used as internal standard.

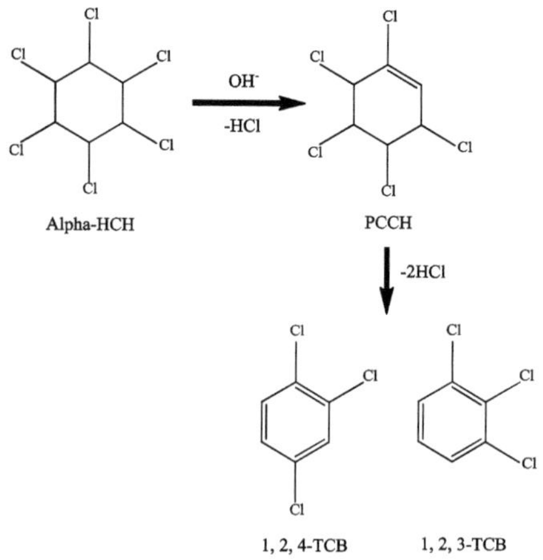

Scheme S3. Proposed reaction mechanism for α-HCH transformation during alkaline hydrolysis at pH 9.78 in Na_2CO_3-$NaHCO_3$ buffer.

Reduction by electrodes or Fe⁰ nanoparticles

Control experiments. The control experiments without electrical current or Fe^0 nanoparticles resulted in insignificant changes in α-HCH concentration. However, a slight trend of α-HCH transformation was found in the control experiment without electric current which may be a result of volatilization.

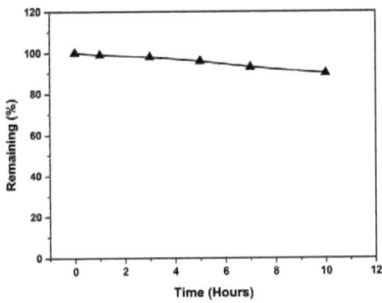

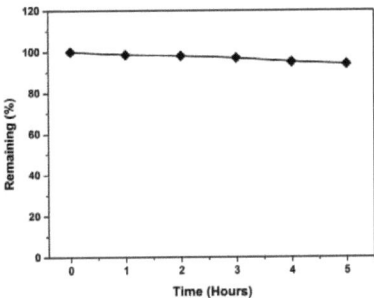

Figure S8. Remaining fraction of α-HCH in the control experiments without electrode (triangle) or Fe^0 nanoparticles (rhombus) in phosphate buffer (pH=7.0).

Products analysis and transformation pathways. The GC-MS spectra obtained for the transformation by electrodes revealed that the same products PCCHa and TeCCH were observed (Figure S9) as in the case of direct photolysis but a different mechanism was proposed (Scheme S4). In the presence of Fe^0 nanoparticles, the reaction yielded TeCCH as the only product identified by GC-MS (Figure S10). Therefore, electrons provided by Fe^0 may attack α-HCH by eliminating two Cl^- from the ring and forming a double bond (Scheme S5).

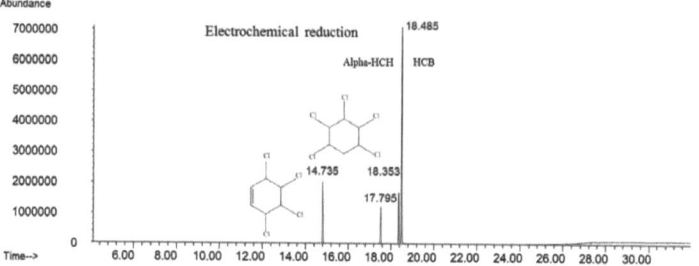

Figure S9. Total ion chromatogram of transformation products formed during electrochemical reduction. HCB was used as internal standard.

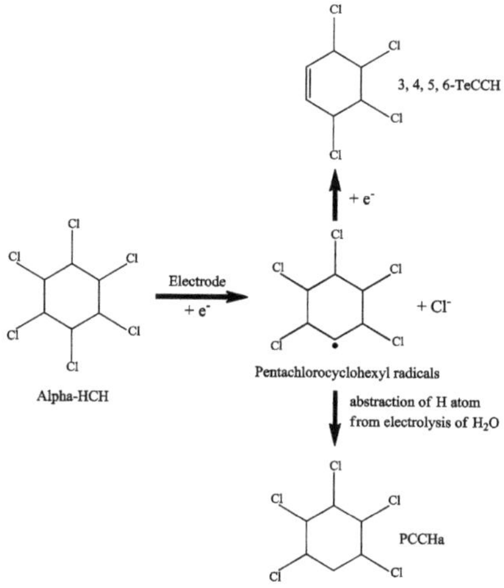

Scheme S4. Proposed reaction mechanism for α-HCH during electrochemical reduction.

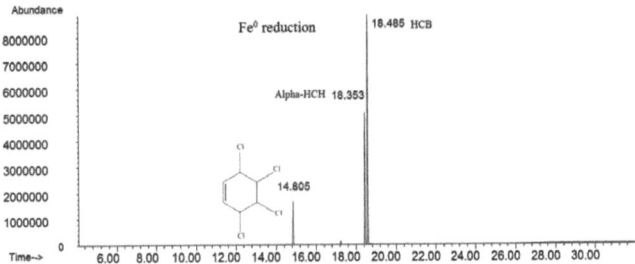

Figure S10. Total ion chromatogram of transformation products formed by α-HCH reduction via Fe⁰ nanoparticles. HCB was used as internal standard.

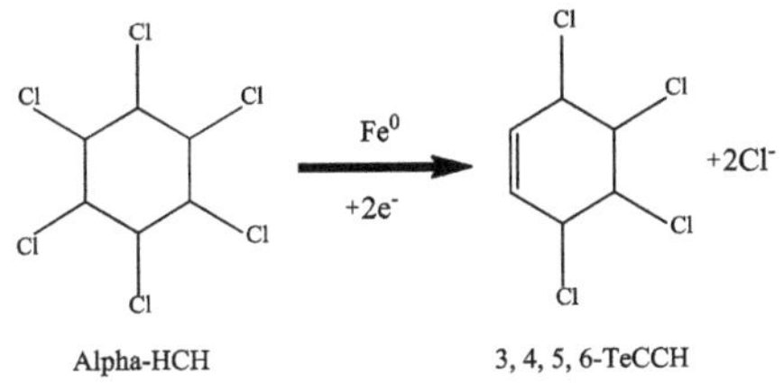

Scheme S5. Proposed reaction mechanism for α-HCH reduction by Fe^0 nanoparticles.

Enantioselectivity

The chiral selectivity is often described as enantiomeric factor (EF). The EF (+) is defined as A$_+$ / (A$_+$ + A$_-$), where A$_+$ and A$_-$ correspond to the peak area or concentrations of (+) and (−) enantiomers (Harner et al., 2000; Vetter and Schurig, 1997). Racemic compounds have an EF (+) equal to 0.5. An EF (+) > 0.5 shows the preferential degradation of (−) enantiomer, and an EF (+) < 0.5 indicates the preferential degradation of (+) enantiomer. EF (−) is defined as A$_-$ / (A$_+$ + A$_-$).

All chemical reaction mechanisms investigated in our study showed no preferential transformation of (+)-enantiomer over (−)-enantiomer (Figure S11) as expected for chemical reactions (Hühnerfuss et al., 1993). In contrast to this finding, microbial biodegradation showed minor to significant enantioselectivity under anoxic and oxic conditions, respectively (Buser and Müller, 1995; Suar et al., 2005). Thus, chemical transformation processes can be distinguished from biotransformation using EF-factors as it was also confirmed by previous studies (Harner et al., 1999; Harner et al., 2000; Jantunen and Bidleman, 1996; Law et al., 2001). However, anaerobic biodegradation of α-HCH exhibited low enantioselectivity (Badea et al., 2011), which might limit the discrimination of abiotic and biological processes using EF-factors.

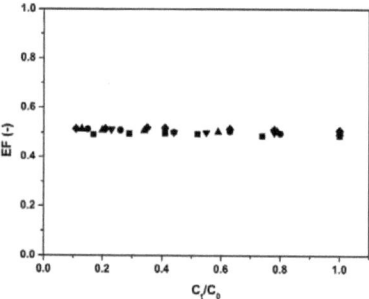

Figure S11. Comparison of concentration decrease (C_t/C_0) vs. EF(-) under direct photolysis (square), indirect photolysis via UV/H_2O_2 (round), alkaline hydrolysis (upward triangle), electrochemical reduction (downward triangle) and reduction by Fe^0 (rhombus).

References

Badea S.L., Vogt C., Gehre M., Fischer A., Danet A.F., Richnow H.H. (2011) Development of an enantiomer-specific stable carbon isotope analysis (ESIA) method for assessing the fate of α-hexachlorocyclo-hexane in the environment. Rapid Communications in Mass Spectrometry, 25(10):1363-1372.

Bashir S., Fischer A., Nijenhuis I., Richnow H.-H. (2013) Enantioselective carbon stable isotope fractionation of hexachlorocyclohexane during aerobic biodegradation by Sphingobium spp. Environmental Science & Technology, 47(20):11432-11439.

Buser H.-R., Mueller M.D. (1995) Isomer and enantioselective degradation of hexachlorocyclohexane isomers in sewage sludge under anaerobic conditions. Environmental Science & Technology, 29(3):664-672.

Clayton G.D., Clayton F.E., Allan R.E., Patty F.A. (1991) Patty's industrial hygiene and toxicology. 4th ed. Wiley, New York.

Coplen T.B. (2011) Guidelines and recommended terms for expression of stable-isotope-ratio and gas-ratio measurement results. Rapid Communication in Mass Spectrometry, 25(17):2538-2560.

Elsner M. (2010) Stable isotope fractionation to investigate natural transformation mechanisms of organic contaminants: principles, prospects and limitations. Journal of Environmental Monitoring, 12(11):2005-2031.

Elsner M., Zwank L., Hunkeler D., Schwarzenbach R.P. (2005) A new concept linking observable stable isotope fractionation to transformation pathways of organic pollutants. Environmental Science & Technology, 39(18):6896-6916.

Elsner M., Cwiertny D.M., Roberts A.L., Lollar B.S. (2007) 1,1,2,2-tetrachloroethane reactions with OH-, Cr(II), granular iron, and a copper-iron bimetal: Insights from product formation and associated carbon isotope fractionation. Environmental Science & Technology, 41(11):4111-4117.

Fiedler H., Hub M., Hutzinger O. (1993) Stoffbericht hexachlorcyclohexan (HCH) LfU.

Harner T., Jantunen L.M.M., Bidleman T.F., Barrie L.A., Kylin H., Strachan W.M.J., Macdonald R.W. (2000) Microbial degradation is a key elimination pathway of hexachlorocyclohexanes from the Arctic Ocean. Geophysical Research Letters, 27(8):1155-1158.

Harner T., Kylin H., Bidleman T.F., Strachan W.M. (1999) Removal of α-and γ-hexachlorocyclohexane and enantiomers of α-hexachlorocyclohexane in the eastern Arctic Ocean. Environmental Science & Technology, 33(8):1157-1164.

Hühnerfuss H., Faller J., Kallenborn R., König W.A., Ludwig P., Pfaffenberger B., Oehme M., Rimkus G. (1993) Enantioselective and nonenantioselective degradation of organic pollutants in the marine ecosystem. Chirality, 5(5):393-399.

Jantunen, L. M., & Bidleman, T. (1996). Air-water gas exchange of hexachlorocyclohexanes (HCHs) and the enantiomers of α-HCH in Arctic regions. Journal of Geophysical Research: Atmospheres (1984–2012), 101(D22), 28837-28846.

Law S.A., Diamond M.L., Helm P.A., Jantunen L.M., Alaee M. (2001) Factors affecting the occurrence and enantiomeric degradation of hexachlorocyclohexane isomers in northern and temperate aquatic systems. Environmental Toxicology and Chemistry 20:2690-2698.

Liu X.M., Peng P.A., Fu J.M., Huang W.L. (2003) Effects of FeS on the transformation kinetics of gamma-hexachlorocyclohexane. Environmental Science & Technology, 37(9):1822-1828.

Mariotti A., Germon J., Hubert P., Kaiser P., Letolle R., Tardieux A., Tardieux P. (1981) Experimental determination of nitrogen kinetic isotope fractionation: some principles; illustration for the denitrification and nitrification processes. Plant and Soil, 62(3):413-430.

Patnaik P. (2011) Handbook of environmental analysis: chemical pollutants in air, water, soil, and solid wastes. CRC Press.

Suar M., Dogra C., Raina V., Kohler H., Poiger T., Hauser A., Buser H., van der Meer J., Holliger C., Lal R. (2005a) Enantioselective transformation of α-HCH by dehydrochlorinase (LinA1, LinA2) from Sphingomonas paucimobilis B90A. Applied and Environmental Microbiology, 71(12):8514-8518.

Vetter, W., & Schurig, V. (1997). Enantioselective determination of chiral organochlorine compounds in biota by gas chromatography on modified cyclodextrins. Journal of Chromatography A, 774(1-2), 143-175.

Supporting Information for chapter 3.4

Evaluating degradation of hexachlorcyclohexane (HCH) isomers within a contaminated aquifer using compound specific stable carbon isotope analysis (CSIA)

Safdar Bashir[†], Kristina L. Hitzfeld[†], Matthias Gehre[†], Hans H. Richnow[†], Anko Fischer[†§]

[†]Helmholtz Centre for Environmental Research - UFZ, Department of Isotope Biogeochemistry, Permoserstr. 15, D-04318 Leipzig, Germany

[§]Isodetect GmbH - Company for isotope monitoring, Deutscher Platz 5b, D-04103 Leipzig, Germany

A 4: Supporting information chapter 3.4

S-1 Standard procedures for groundwater sampling and concentration analyses of pollutants and hydro-geochemical parameters

- Groundwater sampling was performed according to DIN 38402-13 (1985-12) and DVWK-Merkblatt 245/1997 (1997).
- HCHs concentration analysis was performed according to DIN 38407-F2 (1993-02).
- Sulfate und nitrate concentration analysis was performed according to DIN EN ISO 10304-1 D 19 (2009-07).
- Ammonium concentration analysis was performed according to DIN 38406 E 5 (1983-10).
- Methane concentration analysis was performed according to EDI guideline, 2. Part: Surface water (1983).

S-2 CSIA of HCHs

<u>Isotope laboratory standards of HCHs</u>

α-, β-, γ- and δ-HCH were purchased as pure compounds (97-99 %, Sigma-Aldrich Chemie GmbH, Germany) and carbon isotope ratios were determined by elemental analyser – isotope ratio mass spectroscopy (EA-IRMS) (Coplen et al., 2006). The carbon isotope ratios of HCHs were expressed in delta notation ($\delta^{13}C$) vs. V-PDB (see Eq. 1 in the main text of the article) based on a two-point calibration with reference materials obtained from International Atomic Energy Agency (IAEA) (IAEA-CH-6, IAEA-CH-7). The calibration was verified by an additional reference material (IAEA-CH3). HCHs for which carbon isotope ratios were measured by EA-IRMS were used as isotope laboratory standards to monitor the instrument performance of GC-IRMS analyses of HCHs and to check the extraction procedure of HCHs from water samples.

<u>GC-IRMS method parameters</u>

CSIA of HCHs was performed by gas chromatography - isotope ratio mass spectrometry (GC-IRMS) as described previously (Badea et al., 2009). The GC-IRMS system consisted of a gas chromatograph (6890 Series; Agilent Technology, USA) coupled via a GC/C III interface to a MAT 252 mass spectrometer (both Thermo Fisher, Germany). The gas chromatograph was equipped with split/splitless injector and a ZB-1 column (60 m, 0.32 mm, 1 µm; Phenomenex, USA). Helium was used as carrier gas (2 mL/min) for the chromatographic separation. The combustion reactor containing Pt, Ni, CuO (Thermo Fisher, Germany) was operated at 980°C. The combustion oven was re-oxidised frequently and the performance of the combustion was monitored by regular (every six samples) analysis of the isotope laboratory standards of HCHs.

Evaluation of the extraction procedure of HCHs from water phase

In order to evaluate possible isotope effects of the extraction of HCHs from water samples (see section Materials and Methods in the main text) the whole procedure was examined for changes in carbon isotope ratios of HCHs. Glass bottles with a volume of 1150 mL (Schott, Germany) were filled almost completely with tap water. The flasks were spiked with the isotope laboratory standards of HCHs (in acetone) to a final concentration of 100 µg/L for each HCH. Similar to the groundwater samples, 1ml HCl (6M) was added for preservation. The prepared water samples were extracted 3 times with 30 mL DCM in a separating funnel (shaking time 1.5 min). All DCM extracts were combined and dried with anhydrous Na_2SO_4. DCM was evaporated with a rotary evaporator up to 1-2 ml solvent volume. Then the sample was evaporated completelyl in a gentle nitrogen stream at room temperature. The dry extracts were dissolved in 0.5 ml of acetone. The extracts were measured with GC-IRMS and the obtained $\delta^{13}C$-values of HCHs were compared to $\delta^{13}C$-values of HCHs from isotope laboratory standards. All $\delta^{13}C$-values of HCHs obtained for extracts were found to be within the range of ±1 ‰ of $\delta^{13}C$-values of HCHs from isotope laboratory standards (Fig. SI 1). Results suggest that extraction and sample concentration do not lead to significant carbon isotope effects for HCHs.

A 4: Supporting information chapter 3.4

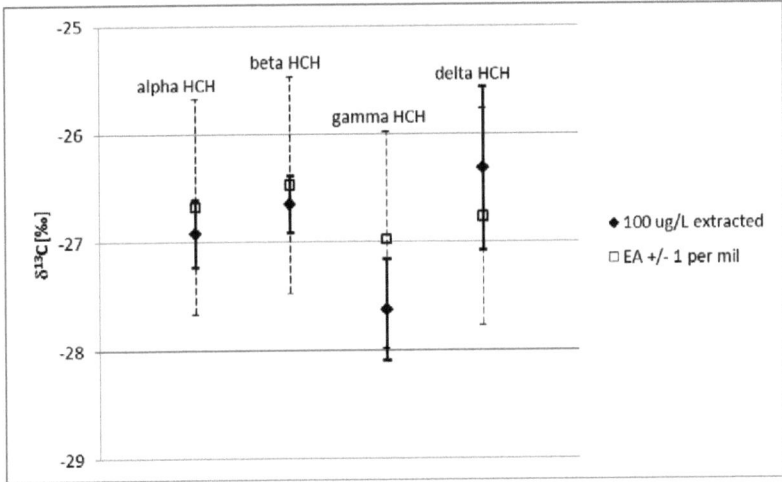

Figure SI 1: Comparison of $\delta^{13}C$-values of HCHs measured after extraction from water and pre-concentration using GC-IRMS to $\delta^{13}C$-values of like HCHs obtained by EA-IRMS.

Method validation

The linearity of the method was evaluated and the limits of detection were defined to improve the sensitivity and precision of the method compared to our previous study (Badea et al., 2009). The linearity range was determined for each HCH isomer to evaluate appropriate injection volumes and the method detection limits of the GC-IRMS system. Mixtures of the isotope laboratory standards of HCHs in acetone at different concentrations were analysed and the results compared to the $\delta^{13}C$-values obtained by EA-IRMS and evaluated with respect to the produced CO_2 signal intensity (m/z 44, in V). Exemplarily, the results for δ-HCH are shown in Fig. SI 2.

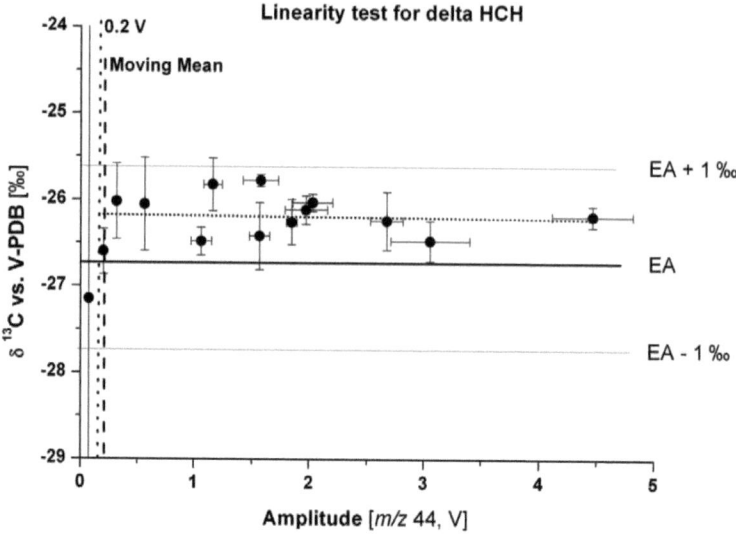

Figure SI 2: Linearity test for GC-IRMS analysis of δ-HCH with method detection limits indicated for arbitrary threshold (vertical dotted line 0.2 V criteria) and moving mean procedure (vertical dashed line). The linear regression for $\delta^{13}C$-value of δ-HCH exceeding the method detection limits (dotted horizontal line) and the $\delta^{13}C$-value obtained by EA-IRMS (black horizontal line) is indicated.

Method detection limits were derived from peak signals, which indicated minimal intensity for accurate and precise $\delta^{13}C$-values. There are different possibilities to determine the lower method detection limit, in this study we choose arbitrary threshold values of an intensity > 200 mV and a standard deviation of 1σ < 0.5 ‰. Another approach is the iterative moving mean procedure for which a moving average of three values has to have a standard deviation of 1σ < 0.5 ‰ (Jochmann et al., 2006). The first average, moving from the most intensive peak of the linearity test to the lower concentration range, which cannot fulfil this criterion, defines the lower detection limit (Jochmann et al., 2006). This latter procedure resulted in higher detection limits, however, very similar linearity slopes and average δ^{13}C-values (Tab. SI 1). Linearity slopes of HCHs were < 0.3 ‰/V (arbitrary thresholds) and even < 0.2 ‰/V for the range of higher intensities (moving mean) (Tab. SI 1). In comparison to previously published linearity (Badea et al., 2009), our approach showed that CSIA with the equipment used allows interpretation of carbon isotope ratios for all HCHs in the concentration range of field samples. Linearity results for δ-HCH, the main HCH isomer of this study, were better than previously published and thus increased reliability for interpretation.

The comparison of results with different intensities was therefore determined to be valid. As field samples were analysed in this study, only results with intensities of at least 200 mV and standard deviations < 1 ‰ were accounted for data interpretation.

Table SI 1: Comparison of method detection limits, average δ^{13}C-values in the linear range and linearity slopes for the two approaches to determine detection limits. [a] Standard deviation of three replicates < 0.2 ‰

Isomer	EA-IRMS δ^{13}C [‰][a]	iterative moving mean procedure (Jochmann et al., 2006)					arbitrary thresholds intensity < 200 mV; σ > 0.5 ‰				
		average δ^{13}C		method detection limit		Linearity slope	Average δ^{13}C		method detection limit		Linearity slope
		[‰]	1σ	min [mV]	max [mV]	[‰/V]	[‰]	1σ	min [mV]	max [mV]	[‰/V]
α	-26.67	-27.1	0.3	756	4089	-0,20	-27.1	0.4	310	4089	-0.25
γ	-26.98	-27.8	0.1	640	4860	-0.01	-27.7	0.3	294	4860	-0.12
β	-26.47	-26.2	0.3	411	3243	0.08	-26.3	0.3	273	3243	0.17
δ	-26.77	-26.2	0.3	203	4466	-0.01	-26.2	0.3	203	4466	-0.01

S-3 Sum concentration of HCHs in 2008

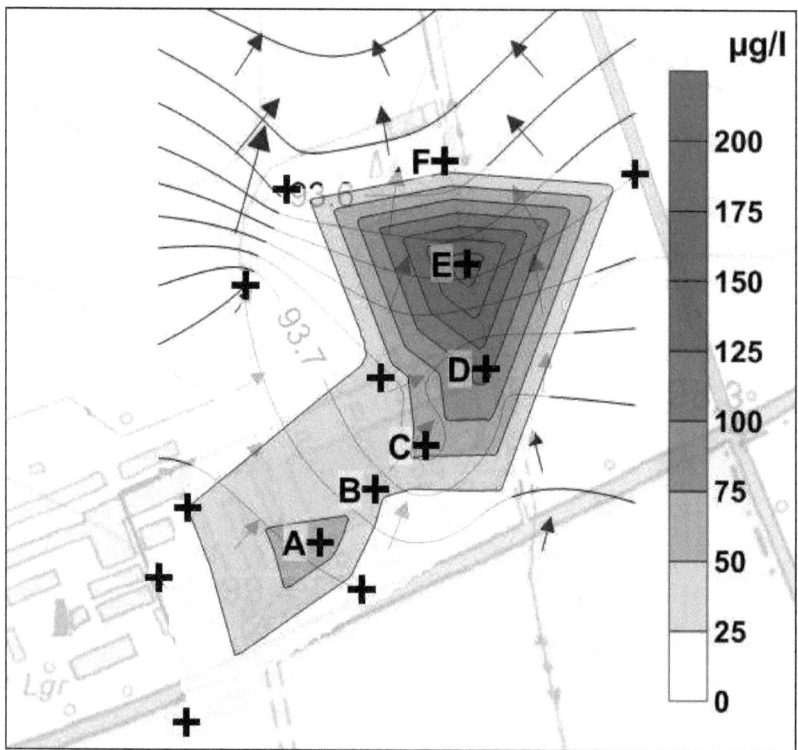

Figure SI 3: Distribution of the sum HCH concentrations [µg/L] and groundwater flow direction (blue arrows) within the upper aquifer of the investigated field site for the monitoring campaign of 2008. Crosses show locations of groundwater wells. Wells A to F represent the investigated aquifer transect within the main groundwater flow direction.

S-4 Hydro-geochemical parameters

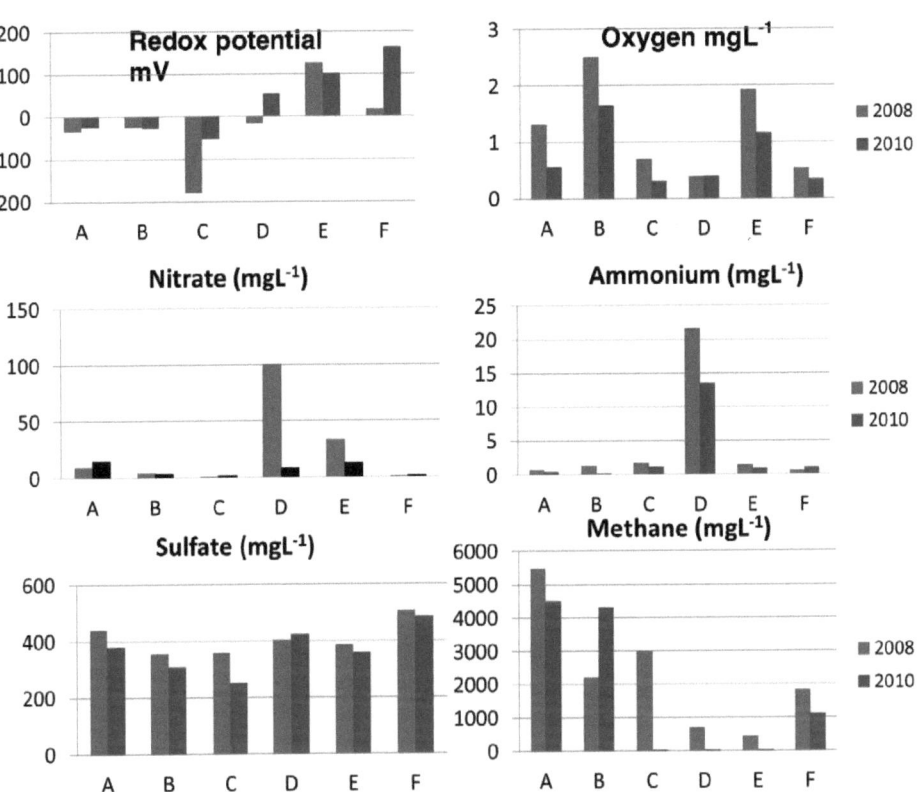

Figure SI 4: Distribution of redox potential, oxygen, nitrate, ammonium, sulfate, and methane concentrations within the investigated transect (wells A to F) along the main groundwater flow direction for 2008 and 2010.

S-5 Qualitative assessment of α-, β- and γ-HCH degradation

α-HCH

The highest concentrations (25 µg/L in 2008; 5.8 µg/L in 2010) and most negative carbon isotope ratios (-29.6‰ in 2008, -28.7 ‰ in 2010) of α-HCH at well E (Fig. 2) confirm the discrete source of the HCH-dump. Lower α-HCH concentrations were found in the source area of the former processing facilities at well A (3.6 µg/L in 2008, 2.5 µg/L in 2010) and at its groundwater downstream at well B (2.3 µg/L in 2008, 1.9 µg/L in 2010) (Fig. 2). However, compared to well E, a similar $\delta^{13}C_{\alpha-HCH}$-value was found at well A (-28.3 ‰) in 2008. In 2010, ^{13}C-enriched α-HCH emerged in wells A (-22.4 ‰) and B (-20.4 ‰). The change from 2008 to 2010 in $\delta^{13}C_{\alpha-HCH}$-values of > +5 ‰ for well A concomitant with decreasing α-HCH concentrations can be attributed to increasing biodegradation of α-HCH in the source area of the former processing facilities. Although the α-HCH concentration decreased at well D from 7.5 µg/L (2008) to 2.1 µg/L (2010), α-HCH became more ^{13}C-depleted (-23.5 ‰ in 2008, -27.2 ‰ in 2010) (Fig. 2). Thus, the decrease in α-HCH concentration in 2010 seemed to be caused primarily by physical processes rather than biodegradation. The overall lower concentrations and more ^{13}C-depleted α-HCH as compared to well E suggest that natural attenuation by physical processes increased while biodegradation decreased at this well.

β-HCH

At well A, β-HCH concentrations in 2008 were lower (4.2 µg/L) than in 2010 (9.3 µg/L) while similar $\delta^{13}C_{\beta-HCH}$-values of -25.4‰ and -25.8 ‰ were determined

(Fig. 2). Thus, the change in concentration can be attributed to physical processes, most probably due to an increased recharge of β-HCH into the groundwater in 2010. Comparably high β-HCH concentrations were found in wells D (9.7 µg/L in 2008, 15 µg/L in 2010) and E (9.5 µg/L in 2008, 5.8 µg/L in 2010) indicating the distinct source zone (the HCH dump) in the area of these wells. Carbon isotope ratios confirmed this finding, showing the most ^{13}C-depleted β-HCH in 2008 in this area (D:-29.8 ‰; E: -27.7 ‰) (Fig. 2). In contrast to the source zone at well A, changes in $\delta^{13}C_{\beta\text{-HCH}}$-values at wells D and E from 2008 to 2010 (shift > +4 ‰) suggest that despite increasing β-HCH concentration at well D, biodegradation proceeded at these wells. Compared to well D, more ^{13}C-enriched β-HCH (+1.8 ‰) and lower β-HCH concentration were observed at well E in 2010, indicating biodegradation of β-HCH between those two wells along the main groundwater flow direction. More significant evidence for biodegradation was provided in the groundwater downstream of the source zone at well A located in the area of the former processing facilities. Compared to well A, a shift in $\delta^{13}C_{\beta\text{-HCH}}$-values by > +7 ‰ could be observed at well B accompanied by a concentration decrease of 8 µg/L, indicating increasing and sustainable biodegradation of β-HCH in this area.

γ-HCH

Concentrations of γ-HCH were very low (< 5 µg/L) in all wells of the investigated transect except for well E (41 µg/L in 2008, 10 µg/L in 2010), thus limiting the evaluation of γ-HCH biodegradation by means of carbon isotope ratios (Fig. 2). High concentrations and consistent, highly ^{13}C-depleted γ-HCH (2008: -29.7‰; 2010: -30.1‰) at well E revealed a discrete HCH source in the area of a former

HCH dump. Similar $\delta^{13}C_{\gamma-HCH}$-values indicate that physical processes seemed to be the main cause of the concentration decrease of γ-HCH at well E from 2008 to 2010. Compared to well E, the lower γ-HCH concentrations and more positive $\delta^{13}C_{\gamma-HCH}$-values at well D (4.5 µg/L, -25.7 ‰) and well A (3.2 µg/L, -24.4 ‰) in 2008 led to the assumption that biodegradation of γ-HCH is occurring at these wells. Overall, carbon isotope ratios and concentrations of γ-HCH seem to correspond with the findings for the other HCH isomers however, due to the low number of carbon isotope ratios available, only limited information could be derived on the biodegradation of γ-HCH.

S-6 References

Richtlinien für die Untersuchung von Abwasser und Oberflächenwasser (Allgemeine Hinweise und Analysemethoden), (1983) Eidgenössisches Departement des Innern (EDI), Switzerland Teil 2

DIN 38406-5, (1983-10) German standard methods for the examination of water, waste water and sludge; cations (group E); determination of ammonia-nitrogen (E 5).

DIN 38402-13, (1985-12) German standard methods for the examination of water, waste water and sludge; general information (group A); sampling from aquifers (A 13).

DIN 38407-2, (1993-02) German standard methods for the determination of water, waste water and sludge; jointly determinable substances (group F); determination of low volatile halogenated hydrocarbons by gas chromatography (F 2).

DVWK-Merkblatt 245/1997, (1997) Tiefenorientierte Probennahme aus Grundwassermessstellen.

ISO 10304-1:2007, (2009-07) Water quality -- Determination of dissolved anions by liquid chromatography of ions -- Part 1: Determination of bromide, chloride, fluoride, nitrate, nitrite, phosphate and sulfate.

Badea S.L., Vogt C., Weber S., Danet A.F., Richnow H.H. (2009) Stable isotope fractionation of gamma-Hexachlorocyclohexane (Lindane) during reductive dechlorination by two strains of sulfate-reducing bacteria. Environmental Science & Technology, 43(9):3155-3161.

Coplen T.B., Brand W.A., Gehre M., Groning M., Meijer H.A.J., Toman B., Verkouteren R.M. (2006) New guidelines for delta C-13 measurements. Analytical Chemistry, 78(7):2439-2441.

Jochmann M.A., Blessing M., Haderlein S.B., Schmidt T.C. (2006) A new approach to determine method detection limits for compound-specific isotope analysis of volatile organic compounds. Rapid Communication in Mass Spectrometry, 20(24):3639-3648.

A 4: Supporting information chapter 3.4

These last three pages left blank intentionally

i want morebooks!

Buy your books fast and straightforward online - at one of world's fastest growing online book stores! Environmentally sound due to Print-on-Demand technologies.

Buy your books online at
www.get-morebooks.com

Kaufen Sie Ihre Bücher schnell und unkompliziert online – auf einer der am schnellsten wachsenden Buchhandelsplattformen weltweit! Dank Print-On-Demand umwelt- und ressourcenschonend produziert.

Bücher schneller online kaufen
www.morebooks.de

VDM Verlagsservicegesellschaft mbH
Heinrich-Böcking-Str. 6-8
D - 66121 Saarbrücken

Telefon: +49 681 3720 174
Telefax: +49 681 3720 1749

info@vdm-vsg.de
www.vdm-vsg.de

Printed by Books on Demand GmbH, Norderstedt / Germany